Concepts and Problems in Physics

Dr. Sanjay Kumar

WORK, ENERGY AND POWER

FOR JEE (MAIN & ADVANCED) & NEET

Dr. Sanjay Kumar

Copyright © 2020 Sanjay Kumar
All rights reserved.
ISBN: 9798681219781

No part of this book may be reproduced or distributed in any form or by any means, electronic, mechanical, photocopying, recording, or otherwise or stored in a database or retrieval system without the prior written permission of the author.

To JEE (MAIN & ADVANCED) & NEET aspirants
With the hope that this work will stimulate
an interest in Physics
and provide an acceptable guide to its understanding.

CONTENTS

1. WORK DONE BY A CONSTANT FORCE 1
1.1. POSITIVE AND NEGATIVE AND ZERO WORKS 1
2. EXPRESSION OF WORK DONE BY A CONSTANT FORCE 1
2.1.1. THE NECESSARY CONDITIONS FOR NON ZERO WORK 2
3. FORCES THAT DO NO WORK 3
4. INTERNAL WORK OF A COMPOSITE SYSTEM (ZERO-WORK SITUATIONS) 3
5. TOTAL WORK 4
6. WORK DONE BY A WEIGHT 4
7. WORK DONE BY FRICTION 5
8. WORK ALONG A CURVED PATH 8
9. CHECKPOINT 1 9
10. WORK DONE BY A VARIABLE FORCE 11
10.1. ONE-DIMENSIONAL ANALYSIS 11
10.2. THREE-DIMENSIONAL ANALYSIS 11
10.3. WORK DONE BY SPRING FORCE 13
11. CHECKPOINT 2 14
12. DEPENDENCE OF WORK ON REFERENCE FRAME 15
13. WORK-ENERGY THEOREM AND DEFINITION OF KINETIC ENERGY 15
13.1. HOW TO APPLY WORK ENERGY THEOREM 16
13.2. WORK AND KINETIC ENERGY IN COMPOSITE SYSTEMS 21
14. CHECKPOINT 3 22
15. WORK OF CONSTRAINT FORCES 23
16. VIRTUAL WORK 25
16.1. VIRTUAL WORK OF A FORCE 25
16.2. VIRTUAL WORK OF A COUPLE 25
16.3. PRINCIPLE OF VIRTUAL WORK 25
17. CONSERVATIVE AND NON-CONSERVATIVE FORCES 25
17.1. CONSERVATIVE (OR INTERNAL) FORCES 26
17.2. NON-CONSERVATIVE (OR EXTERNAL) FORCES 27
17.3. IDENTIFICATION OF A PLANAR CONSERVATIVE FORCE 27
18. POTENTIAL ENERGY 28
18.1. GRAVITATIONAL POTENTIAL ENERGY 29
18.1.1. GRAVITATIONAL POTENTIAL ENERGY WHEN GRAVITATIONAL FORCE IS UNIFORM 29
18.1.2. GRAVITATIONAL POTENTIAL ENERGY WHEN GRAVITATIONAL FORCE IS NON-UNIFORM 31
18.2. ELASTIC POTENTIAL ENERGY 33
18.2.1. GRAVITATIONAL POTENTIAL ENERGY VS. ELASTIC POTENTIAL ENERGY. 33
19. CHECKPOINT 3 34
20. MECHANICAL ENERGY 35
20.1. POSITIVE VS. NEGATIVE WORK AND ENERGY CHANGE 35
21. CONSERVATION OF MECHANICAL ENERGY 36
22. NON-MECHANICAL ENERGY 37
23. GENERAL FORM OF CONSERVATION OF ENERGY 37
24. WORK DONE IN MAKING A PYRAMID OF SAND 39
25. CONSERVATIVE FORCE AS A NEGATIVE GRADIENT OF POTENTIAL ENERGY 39
25.1. IN ONE DIMENSION 39
25.2. IN THREE DIMENSIONS 40
26. ENERGY DIAGRAMS 40
26.1. THE POTENTIAL ENERGY CURVE 40
27. POWER 43
27.1. PHYSICAL SIGNIFICANCE OF POWER 44
27.2. MEANING OF A 60 W BULB 45
28. EFFICIENCY 45
29. SOME IMPORTANT CONVERSION FACTORS 45
30. CHECKPOINT 4 47
31. SOLVED EXAMPLES 48
32. QUESTIONS AND EXERCISES 53
32.1. CONCEPTUAL QUESTIONS 53
32.2. PROBLEMS 55
32.3. MULTIPLE CHOICE PROBLEMS 60
32.4. MULTIPLE CHOICE ASSIGNMENTS 65
32.4.1. LEVEL 1 65
32.4.1.1. WORK 65
32.4.1.2. POWER 66
32.4.1.3. KINETIC ENERGY 67
32.4.1.4. POTENTIAL ENERGY 67
32.4.1.5. CONSERVATION OF MECHANICAL ENERGY 67
32.4.2. LEVEL 2 68
32.4.3. LEVEL 3 69
32.4.4. LEVEL 4 71
33. ANSWERS KEYS AND SOLUTIONS 75
33.1. CHECKPOINT 1 75
33.2. CHECKPOINT 2 78
33.3. CHECKPOINT 3 78
33.4. CHECKPOINT 3 80
33.5. CHECKPOINT 4 80
33.6. CONCEPTUAL QUESTIONS 83
33.7. PROBLEMS 87
33.8. MULTIPLE CHOICE PROBLEMS 97
33.9. MULTIPLE CHOICE ASSIGNMENTS 103

33.9.1. LEVEL 1   103
33.9.2. LEVEL 2   103
33.9.3. LEVEL 3   103
33.9.4. LEVEL 4   103

33.9.4.1. SECTION A   103
33.9.4.2. SECTION B   103

PREFACE

This physics book is the product of more than fifteen years of teaching and innovation experience in physics for JEE (Main & Advanced) and Medical aspirants. Our main goals in writing this book are

- to present the basic concepts and principles of physics that students need to know for JEE-advanced and other related competitive exams.
- to provide a balance of quantitative reasoning and conceptual understanding, with special attention to concepts that have been causing difficulties to student in understanding the concepts.
- to develop students' problem-solving skills and confidence in a systematic manner.
- to motivate students by integrating real-world examples that build upon their everyday experiences.

What's New?

Lots! Much is new and unseen before. Here are the big four:

1. Every concept is given in student friendly language with various solved problems. The solution is provided with problem solving approach and discussion.
2. Checkpoint questions have been added to applicable sections of the text to allow students to pause and test their understanding of the concept explored within the current section. The answers to the Checkpoints are given in answer keys, at the end of the chapter, so that students can confirm their knowledge without jumping too quickly to the provided answer.
3. Special attention is given to all tricky topics (like- work energy theorem, conservative and non conservative forces, conservation of mechanical energy, work done by non conservative forces, power of pump and chain related problems), so that student can easily solve them with fun.
4. To test the understanding level of students, multiple choice questions, conceptual questions, practice problems with previous years JEE Main and Advanced problems are provided at the end of the whole discussion. Number of dots indicates level of problem difficulty. Straightforward problems (basic level) are indicated by single dot (●), intermediate problems (JEE mains level) are indicated by double dots (●●), whereas challenging problems (advanced level) are indicated by thee dots (●●●). Answer keys with hints and solutions are provided at the end of the chapter.

We have kept these goals in mind while developing the main themes of our physics book.

Dr. Sanjay Kumar

This page intentionally left blank

1 WORK, ENERGY AND POWER

1. WORK DONE BY A CONSTANT FORCE

The word w*or*k has a variety of meanings in everyday language such as- studying for an exam, carrying a backpack, or riding a bike. But in physics, work is the transfer of energy by a force.

1.1. POSITIVE AND NEGATIVE AND ZERO WORKS

Work W is the energy transferred to or from an object by means of an external force acting on the object. Energy transferred to the object is a positive work done by the external force (while it is a negative work done by the system force), and energy transferred from the object is negative work done by an external force (while it is a positive work done by the system force).

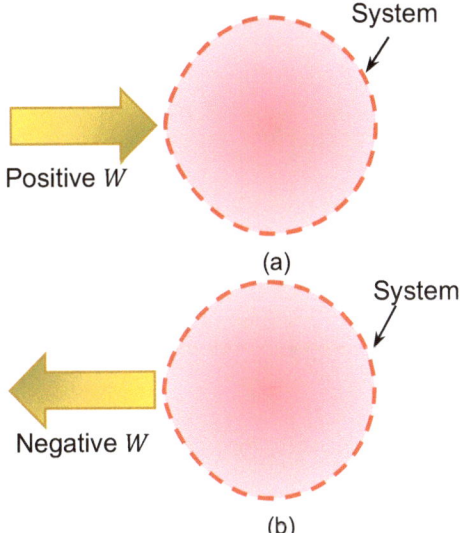

Fig. 1. (a) Positive work W done on an arbitrary system means a transfer of energy to the system. (b) Negative work W means a transfer of energy from the system.

Work is a scalar quantity that can be positive, negative, or zero. The work done by object A on object B is positive if energy is transferred from A to B, and is negative if energy is transferred from B to A. If no energy is transferred, the work done is zero.

The term *transfer* can be misleading. It does not mean that anything material flows into or out of the object; that is, the transfer is not like a flow of water. Rather, it is like the electronic transfer of money between two bank accounts: The number in one account goes up while the number in the other account goes down, with nothing material passing between the two accounts.

2. EXPRESSION OF WORK DONE BY A CONSTANT FORCE

In physics, work is done on an object by a force when the point of application of the force moves through a displacement. Work done by a constant force *is defined as the product of the continuously attached component of the force in the direction of the displacement and the magnitude of the displacement of point of application of force.*

To understand the concept clearly, let us consider that a bead slides without friction along a thin horizontal rod. The bead moves from A to B, which we represent by the displacement vector $\vec{s}$. A constant force $\vec{F}$ is exerted on the bead at an angle θ, with horizontal direction (Fig. 1), by an external agent;. The components of $\vec{F}$ along the displacement vector is $F\cos\theta$ whereas perpendicular to the displacement vector, the component of force is $F\sin\theta$.

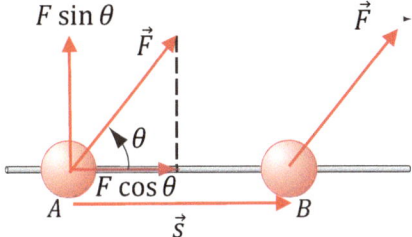

FIFURE 1. A bead slides along a thin rod from A to B. A constant force $\vec{F}$ which makes an angle θ with the wire, acts on the bead at every point between A and B.

Only the component of the force $F\cos\theta$ along the displacement vector $\vec{s}$ contributes to the work, so the work done by the force $\vec{F}$ is,

$$W = (F\cos\theta)s = Fs\cos\theta \qquad \ldots (1)$$
(constant force)

Equation 1 gives the work done by the particular force $\vec{F}$.

Eq. (1), can also be written as the scalar product of force and displacement vectors, such as-

$$W = \vec{F}\cdot\vec{s} \qquad \ldots (2)$$

In above example, just for simplicity, we have considered a single force $\vec{F}$. But there may be several forces acting on the object; for example, in Fig. 2, in addition to the force $\vec{F}$ there is the normal force $\vec{N}$, the force of gravity $m\vec{g}$, and also a frictional force $\vec{f}_k$. We must calculate the work separately for each force that acts on the object.

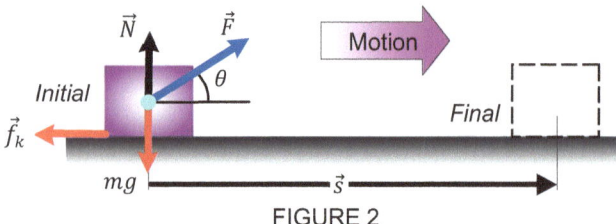

FIGURE 2

The unit of work in SI units is $N.m$ [abbreviated by joule (J)], i.e., $1J = 1N.m$, and in CGS units is $dyne.cm$ (abbreviated by erg), i.e. $1\,erg = 1\,dyne.cm$. Note that $1\,joule = 10^7\,erg$

System	Unit of work	Name of combined unit
SI	$N.m$	joule (J)
CGS	$dyne.cm$	erg
British	$ft.lb$	$ft.lb$

The equation (1), can also be written as-

$$W = Fs\cos\theta = F(s\cos\theta)$$

Magnitude of force | Component of displacement in the direction of the force

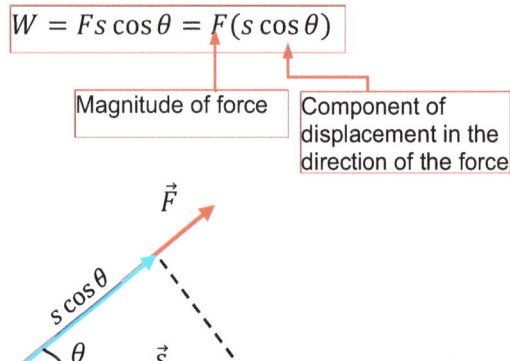

FIGURE 3

Therefore, we can also define the work done by a force as- *the component of displacement (of point of application of force), in the direction of the force times the magnitude of the force.*

Definitions of Work in Physics

The work done by a force can be thought of in the following two equivalent ways:
1. Work is the component of force in the direction of the displacement times the magnitude of the displacement of point of application of force
2. Work is the component of displacement (*of point of application of force*) in the direction of the force times the magnitude of the force.

➤ Work depends on the angle between the force, $\vec{F}$, and the displacement (or direction of motion), $\vec{s}$.

As $\cos\theta = \begin{cases} +ve, if -90° < \theta < 90° \\ 0, \quad if\ \theta = \pm 90° \\ -ve, if\ 90° < \theta < 270° \end{cases}$

Therefore,

$W = (F\cos\theta)s = \begin{cases} +ve, if -90° < \theta < 90° \\ 0, \quad if\ \theta = \pm 90° \\ -ve, if\ 90° < \theta < 270° \end{cases}$

These, three distinct possibilities, are shown below-

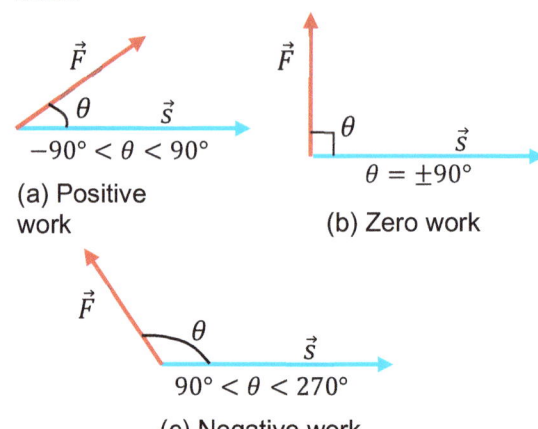

(a) Positive work
(b) Zero work
(c) Negative work

FIGURE. 4

✓ Work is positive if the force has a component in the direction of motion of its point of application.
$(-90° < \theta < 90°)$
✓ Work is zero if the force has no component in the direction of motion of its point of application. $(\theta = \pm 90°)$
✓ Work is negative if the force has a component opposite to the direction of motion of its point of application.
$(90° < \theta < 270°)$

Thus, whenever we calculate work, we must be careful about its sign and not just assume it to be positive.

In Fig. 2, the gravitational force on the block i.e., the weight and the normal reaction both are perpendicular to the displacement, therefore, the work done by these two forces on the block is zero.
As, the force of friction is opposite the displacement vector, therefore from-
$W = Fs\cos\theta$, we get
$W = f_k s \cos 180° = -f_k s \quad (\because \cos 180° = -1)$
Thus, the work done by force of friction is negative.

2.1.1. THE NECESSARY CONDITIONS FOR NON ZERO WORK

It is important to note that-for a non zero work by a force on an object-
1. The force should continuously be attached at the point of application of the given object.
2. There should always be a displacement of the point of application of force on the given body.

In case of a rigid body, every point of it moves with same displacement. Therefore, we generally speak the displacement of the body not displacement of point of application of the force.

3. FORCES THAT DO NO WORK

If the angle between force vector and the displacement of point of application is 90°, then the work done by the force will be zero. The work will also be zero, if either there is no continuously attached component of force on the object or the displacement of point of application of the force is zero.

For example, no work is done by the tension in the string on a swinging pendulum bob because the tension is always perpendicular to the velocity of the bob (Fig. 1a). Similarly, no work is done by the Earth's gravitational force on a satellite in circular orbit (Fig. 1b). In a circular orbit, the gravitational force is always directed along a radius from the satellite to the center of the Earth. At every point in the orbit, the gravitational force is perpendicular to the velocity of the satellite (which is tangent to the circular orbit).

By contrast, gravity does work on a satellite in a noncircular orbit (Fig. 1c). Only at points A and P are the gravitational force and the satellite's velocity perpendicular. Wherever the angle between the gravitational force and the velocity is less than 90°, gravity is doing positive work, increasing the satellite's kinetic energy by making it move faster. Wherever the angle between the gravitational force and the velocity is greater than 90°, gravity is doing negative work, decreasing the satellite's kinetic energy by slowing it down.

4. INTERNAL WORK OF A COMPOSITE SYSTEM (ZERO-WORK SITUATIONS)

A human body is an example of a composite system of various muscles. As, different parts of a human body have different motions, therefore, these muscles always works on each other when they .contract and expand. Now, consider some special situations shown in Fig. 1-

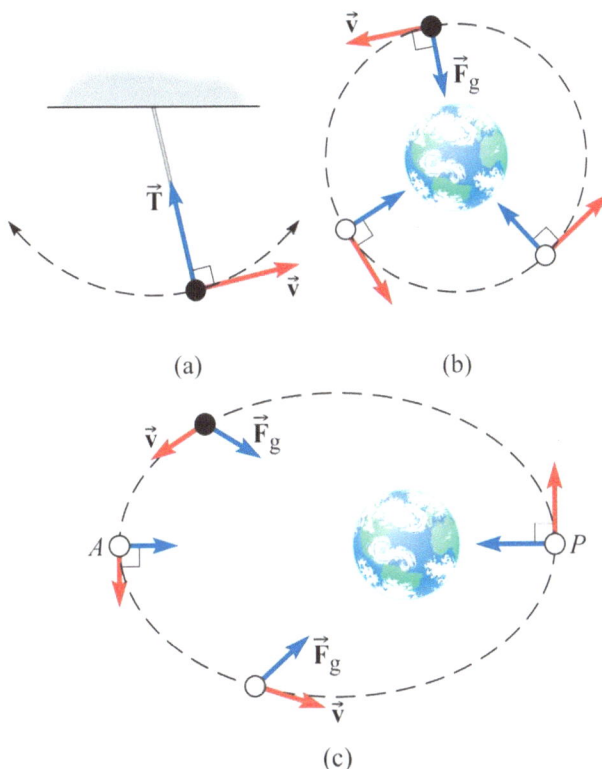

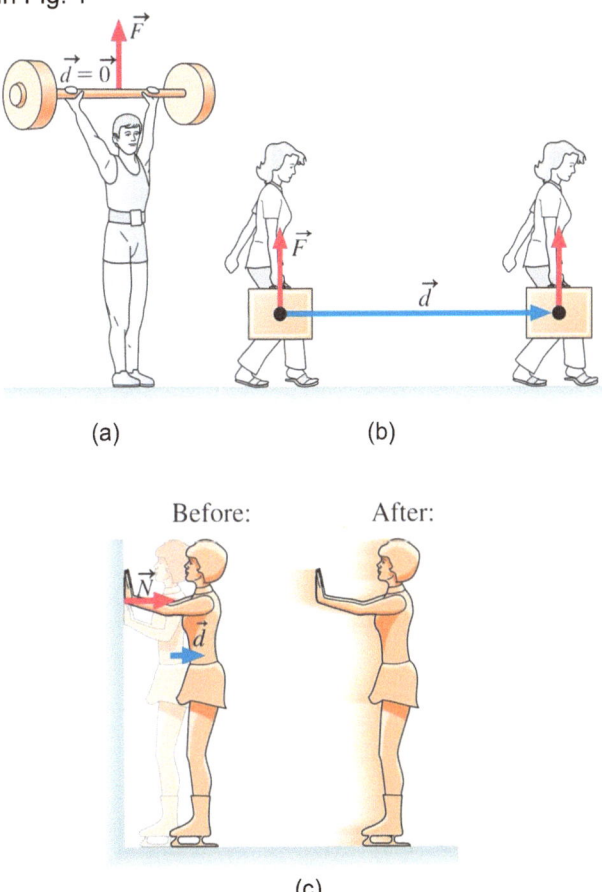

FIGURE 1 (a) The tension in the string of a pendulum is always perpendicular to the velocity of the pendulum bob, so the string does no work on the bob. (b) No matter where the satellite is in its circular orbit, it experiences a gravitational force directed toward the center of the Earth. This force is always perpendicular to the satellite's velocity; thus, gravity does no work on the satellite. (c) In an elliptical orbit, the gravitational force is *not* always perpendicular to the velocity. As the satellite moves counterclockwise in its orbit from point P to point A, gravity does negative work; from A to P, gravity does positive work.

FIGURE 1

Figure 1(a): If the point of application of force undergoes no displacement while the force acts, no work is done. This can sometimes seem counterintuitive. The weightlifter struggles greatly to hold the barbell over his head. But during the time the barbell remains stationary, he does no work on it because its displacement is zero.

Why, then, does the weightlifter become tired and eventually lose his ability to support the weights? If we examine his muscles, we find that work is being done microscopically even when the weight does not move. A muscle is not a rigid support and cannot sustain a load in a static manner. The individual muscle fibers repeatedly relax and contract, and work is done in each contraction. This microscopic work drains his internal supply of energy, and gradually he becomes too tired to hold the weights.

In this chapter we do not consider this "internal" form of work. We use *work* only in the strict sense of $W = Fs\cos\theta$, so that it does indeed vanish when there is no motion of the point of application of force of the body

Figure 1(b): A force perpendicular to the displacement does no work. The woman exerts only a vertical force on the briefcase she's carrying. This force has no component in the direction of the displacement, so the briefcase moves at a constant velocity and its kinetic energy remains constant. Since the energy of the briefcase doesn't change, it must be that no energy is being transferred to it as work. (This is the case where $\theta = 90°$ in $W = Fs\cos\theta$).

Note that, in this case, the work done by force of friction between shoes and ground is also zero because, when the shoes remains in contact with the ground, there is no displacement of the point of application of force and as soon as the woman raises her legs to move forward, the force of friction between her shoes and ground becomes zero. The reason why she gets tired is that, it's not adequate to represent the woman as a single point mass. Different parts of the woman's body have different motions; The various parts of her body interact with each other, and one part can exert forces and do work on another part.

Figure 1(c): If the point of application of force undergoes no displacement, no work is done. Even though the wall pushes on the skater with a normal force $\vec{N}$ and she undergoes a displacement $\vec{d}$, the wall does no work on her, because the point of her body on which $\vec{N}$ acts—her hands—undergoes no displacement.

> **KEYPOINT** Work, unlike properties such as mass, volume, or temperature, is not an intrinsic property of a body. We cannot say, for example, that a body gains, loses, or contains a certain amount of work when it moves through a distance as a force acts on it. Work is associated with the force that acts on the body, or with the agent that exerts that force.

5. TOTAL WORK

When more than one force acts on an object, the total work is the sum of the work done by each force separately. Thus, if force $\vec{F}_1$ does work W_1, force $\vec{F}_2$ does work W_2 and so on, the total work is

$$W_{total} = W_1 + W_2 + W_3 + \cdots = \sum W$$

Equivalently, the total work can be calculated by first performing a vector sum of all the forces acting on an object to obtain $\vec{F}_{total}$ and then using our basic definition of work

$$W_{total} = (F_{total}\cos\theta)s = F_{total}s\cos\theta$$

where θ is the angle between $\vec{F}_{total}$ and the displacement $\vec{s}$.

6. WORK DONE BY A WEIGHT

Consider a block of mass m to be lifted up with almost zero acceleration (i.e., $\vec{a} \simeq 0$) by a constant force $\vec{F}$ applied by a person, see Fig. 1. While in motion, the force $\vec{F}$ and the weight $m\vec{g}$ will be oppositely directed but equal in magnitude, i.e.

$$\vec{F} = m\vec{g} \qquad \ldots (2)$$

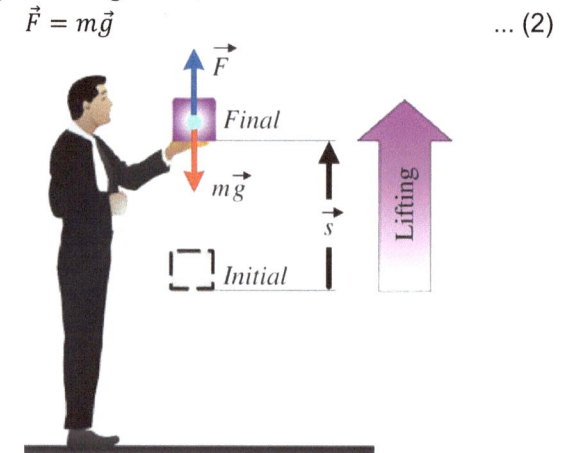

FIGURE. 1 Lifting block with almost zero acceleration

If the upward displacement of the block is denoted by $\vec{s}$, as in Fig. 1, then we can calculate the work done by $\vec{F}$ as follows:

$$W_F = \vec{F}\cdot\vec{s} = Fs\cos 0° = Fs = mgs \qquad \ldots (3)$$

where, we have used the fact that the angle between the two parallel vectors $\vec{F}$ and $\vec{s}$ is zero.

Also, we can calculate the work done by the gravitational force $m\vec{g}$ as follows:

$$W_g = m\vec{g}\cdot\vec{s} = mgs\cos 180° = -mgs \qquad \ldots (4)$$

Thus, we conclude that:

$W_F = mgs$ and $W_g = -mgs$ (Lifting case) ... (5)

where we have used the fact that the angle between the two antiparallel vectors $m\vec{g}$ and $\vec{s}$ is $180°$. The net work $W_F + W_g$ done on the block is zero, as expected, because the net force on the block is zero. This is not, of course, to say that it takes no work to lift a block through a vertical height s. In such a context, we do not refer to the net work, but to the work done by the person.

When we lower the block vertically downward with almost zero acceleration for a displacement $\vec{s}$, see

Fig.2, the sign of the work done by $\vec{F}$ and $m\vec{g}$ will be reversed, since the sign of $\vec{s}$ has reversed.
Following similar steps, one can easily find:
$W_F = \vec{F}.\vec{s} = Fs \cos 180° = -Fs = -mgs$ ... (6)
$W_g = m\vec{g}.\vec{s} = mgs \cos 0° = mgs$ ... (7)
Thus, we conclude that:
$W_F = -mgs$ and $W_g = mgs$ (Lowering case) ... (8)

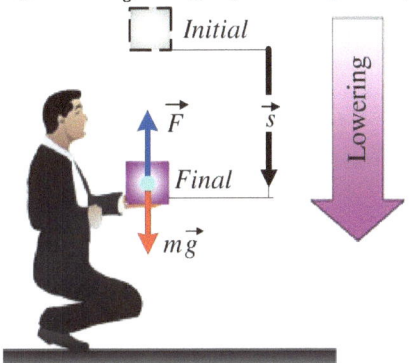

FIGURE. 2 Lowering down a block with almost zero acceleration

7. WORK DONE BY FRICTION

A common example in which the work is always negative is the work done by friction. When a block slides over a rough surface due to an applied force $\vec{F}$, as shown in Fig. 1, the work done by the frictional force $\vec{f_k}$ while the block undergoes a displacement $\vec{s}$ is:
$W_f = \vec{f_k}.\vec{s} = f_k s \cos 180° = -f_k s$ ... (9)

FIGURE 1

FIGURE. 1 The work done by the kinetic frictional force $\vec{f_k}$ while the block undergoes a displacement $\vec{s}$ is always negative and equals $W_f = -f_k s$

From Fig. 1, one can easily find the work done by gravity, the normal force, and the applied force as follows:
$W_g = m\vec{g}.\vec{s} = mgs \cos 90° = 0$ ... (10)
$W_N = \vec{N}.\vec{s} = Ns \cos 90 = 0$ ... (11)
$W_F = \vec{F}.\vec{s} = Fs \cos \theta$ ... (12)
Here, $0 < \theta < 90°$, therefore $W_F = +ve$
(i) We can note that work is a scalar quantity.
(ii) $dW = |\vec{F}||d\vec{S}| \cos \theta$ i.e. if component of force is along displacement ($\theta < 90°$) work is positive otherwise work is negative ($\theta > 90°$).

EXAMPLE 1. A man rowing a boat upstream is at rest with respect to the shore. (a) Is he doing any work? (b) If he stops rowing and moves down with the stream, is any work being done on him?

APPROACH Work is defined as the product of component of force in the direction of displacement and displacement of the point of application of force.
ANSWER (a) As the displacement of the boat relative to the shore is zero (static equilibrium), so work done by his force (or force of flow of stream or by net force) is zero, i.e., he is doing no work (though he is applying a force).
(b) If he stops rowing, the force of flow will produce displacement relative to the shore in the direction of flow of water. So, work will be done by the force of flowing water and will be positive (and so, kinetic energy will increase).

EXAMPLE 2. Mountain roads rarely go straight up the slope but wind up gradually. Explain why?
ANSWER As gravitational field is conservative, the work done in a gravitational field is path independent, i.e., the work done in going from bottom to the top of a hill is the same whatever be the path followed, i.e.,
W = constant
or $\vec{F}.\vec{s}$ = constant (assuming, $\theta = 0°$)
So, if we increase the length of the path by decreasing the slope, the force needed will be small. On the other hand, if we decrease the length of the path by increasing the slope, the force needed to do the same work will be large. However, the magnitude of force that can be applied by man or machine is limited. So, it is more practical and economical to increase the length of the path rather than force to do a given amount of work. Actually, the increase in path length dilutes the force needed. This is why the mountain roads, generally wind up gradually and rarely go up straight.

☞ *Dissipative forces (like friction) always increase the thermal energy; they never decrease it.*

EXAMPLE 3 (a) Determine the work a hiker must do on a 15.0-kg backpack to carry it up a hill of height $h = 10.0 \, m$, as shown in Fig. 1. Determine also (b) the work done by gravity on the backpack, and (c) the net work done on the backpack. For simplicity, assume the motion is smooth and at constant velocity (i.e., acceleration is zero.

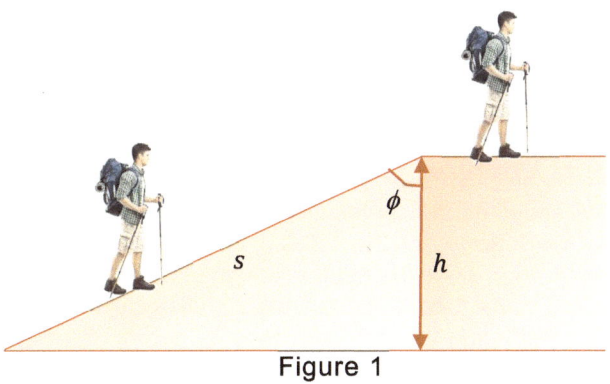

Figure 1

APPROACH Work done, corresponding to each force, is defined as the dot product of the force and displacement of point of application of force, i.e.,
$$W = \vec{F}.\vec{s} = Fs\cos\theta \qquad ...(1)$$
To determine the work corresponding to each force, we have to find each individual force, and the angle, θ, between force and displacement vector.
SOLUTION Newton's second law applied in the vertical direction to the backpack gives-

Figure 2 *FBD* of backpack

$\Sigma F_y = ma_y$
or $F_H - mg = 0$, since, $a_y = 0$
or $F_H = mg = (15.0\text{kg})(9.80\text{m/s}^2) = 147\text{N}$

Work done by a specific force. (*a*) From geometry of Fig.1, the angle between force vector and displacements vector, $\theta = \phi$, therefore from equation (1), the work done by hiker

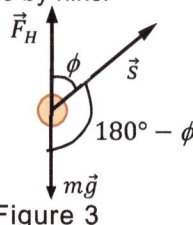

Figure 3

$W_H = F_H s\cos\phi = F_H h \quad (\because \text{from Fig.1, } h = s\cos\phi)$
$= (mg)h]$

From above expression, it is clear that the work done depends only on the change in elevation and not on the angle of the hill, θ. The hiker would do the same work to lift the pack vertically the same height h.
Substituting the given values in above equation, we get
$$W_H = mgh = (147\text{N})(10.0\text{m}) = 1470$$

(*b*) The work done by gravity on the backpack is (from Eq. 1 and Fig. 3)
, $W_G = F_G s\cos(180° - \phi)$
Since, $\cos(180° - \phi) = -\cos\phi$, we have
$W_G = F_G s(-\cos\phi) = mg(-s\cos\phi) = -mgh$
$= -(15.0\text{kg})(9.80 \ m/s^2)(10.0\text{m}) = -1470\ \text{J}$

NOTE The work done by gravity (which is negative here) doesn't depend on the angle of the incline, only on the vertical height h of the hill. This is because gravity acts vertically, so only the vertical component of displacement contributes to work done.
Net work done. (*c*) The *net* work done on the backpack is $W_{net} = 0$, since the net force on the backpack is zero (it is assumed not to accelerate significantly). We can also determine the net work done by adding the work done by each force:
$$W_{net} = W_G + W_H = -1470\text{J} + 1470\text{J} = 0$$

NOTE Even though the *net* work done by all the forces on the backpack is zero, the hiker *does do* work on the backpack equal to 1470 J

EXAMPLE 4. A block of ice is drawn through a distance 500 *cm* along a smooth horizontal surface. The pull in the rope in 100 dyne and the angle between the rope and ground in 30°. Find the work done.
APPROACH Work done is defined as the dot product of force and displacement of point of application of force, i.e.,
$$W = \vec{F}.\vec{s} = Fs\cos\theta \qquad ...(1)$$
SOLUTION Substituting the given values in equation (1), we get
$W = F\cos\theta \times s = 100 \times 500 \ \cos 30°$
$= 43300 \ erg.$

EXAMPLE 5. *A block of mass* 2 kg *is lying on a flat car, which is accelerating with constant acceleration of* $1 \ m/s^2$. *Find the work done by friction on the block in* 10 *m journey of car*
(A) *with respect to driver of car and*
(B) *with respect to a ground observer.*
APPROACH Since displacement depends on reference frame, therefore, work also depends on the reference frame.
Apply, $W = \vec{F}.\vec{s} = Fs\cos\theta$, in each frame.
SOLUTION (A) With respect to driver, the displacement of block is zero therefore, work done by friction force on block is zero.
(B) With respect to ground observer, displacement is $10\ m$ and the force of static friction which is accelerating the block, is
$$f_s = ma = 2 \times 1 = 2N$$
∴ Work done by friction
$$W = \vec{f_s}.\vec{s} = 2 \times 10 = 20J$$
EXAMPLE 6. Find work done by gravity and normal reaction, when block comes from A to B.

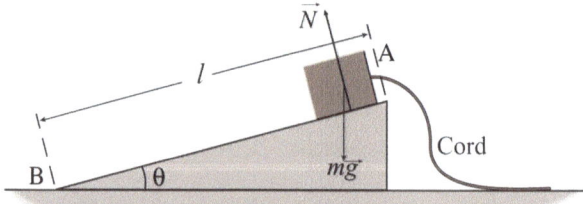

APPROACH Apply work equation, $W = \vec{F}.\vec{s} = Fs\cos\theta$, corresponding to each force.

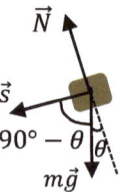

SOLUTION Work done by gravity:
The gravitational force on the block $F = mg$
The displacement, $s = l$,

The angle between $m\vec{g}$ and displacement vector is $90 - \theta$, therefore, the work done by gravity
$$W_{grav} = mgl \cos(90° - \theta)$$
$$= mgl \sin\theta$$
Work done by normal reaction: As the angle between normal reaction $\vec{N}$ and the displacement vector is 90°, therefore
$$W_N = 0$$

EXAMPLE 7. A $10\,kg$ block placed on a rough horizontal floor is being pulled by a constant force $50\,N$. Coefficient of kinetic friction between the block and the floor is 0.4. Find work done by each individual force acting on the block over displacement of $5\,m$ (take $g = 10\,m/s^2$).

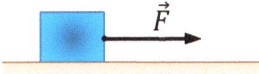

APPROACH The forces acting on the block are-

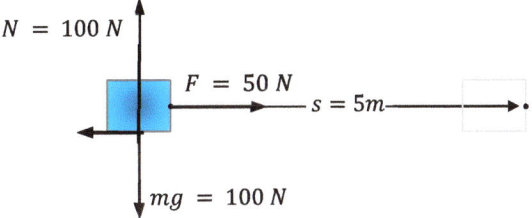

1. Applied pulling force $\vec{F}$: acting parallel to displacement vector, i.e., along positive direction of x axis.
   Given that $F = 50$ newton
2. Gravitational force $M\vec{g}$: acting in vertically downward direction
   $$Mg = 10g = 100 \text{ newton}$$
3. Normal reaction $\vec{N}$: acting in vertically upward direction.
   $$N = Mg = 100 \text{ newton}$$
4. Force of kinetic friction (f_k): acting opposite to displacement vector i.e., along negative direction of x axis.
   Force of kinetic friction,
   $$f_k = \mu_k N = (0.4)(10g) = 40 \text{ newton}$$

As all above forces are constant, therefore, for each force use the work formula-
$$W = \vec{F}.\vec{s} = Fs\cos\theta$$
here, θ is the angle between force $\vec{F}$ and displacement $\vec{s}$ of point of application of the forec.

SOLUTION 1. Work done by applied pulling force:
$$W_1 = Fs\cos\theta = (50 \text{ newton})(5m)\cos 0$$
$$= 250\,J$$
2. Work done by the gravitational force:
$$W_2 = (Mg)s\cos 90°$$
$$= (100 \text{ newton})(5\,m)(0) = 0\,J$$
3. Work done by the normal reaction:
$$W_3 = (N)s\cos 90°$$
$$= (100 \text{ newton})(5\,m)(0) = 0\,J$$
4. Work done by force of kinetic friction:
$$W_4 = f_k s \cos\theta = (40 \text{ newton})(5m)\cos 180°$$
$$= -200\,J$$

EXAMPLE 8. A 10 kg block placed on a rough horizontal floor is being pulled by a constant force $100\,N$ acting at angle 37°. Coefficient of kinetic friction between the block and the floor is 0.4. Find work done by each individual force acting on the block over displacement of 5 m.

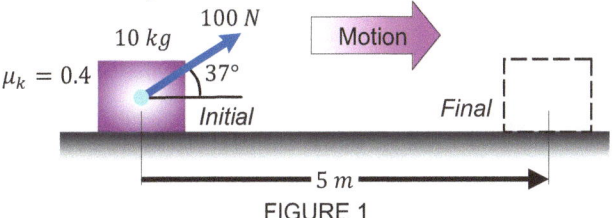
FIGURE 1

APPROACH The forces acting on the block are-

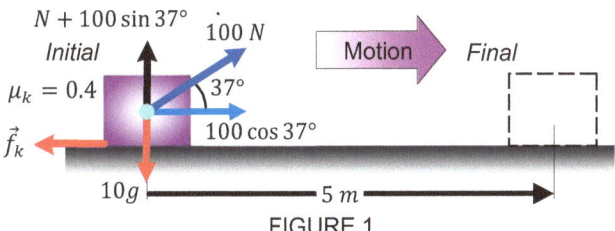
FIGURE 1

1. Applied pulling force $\vec{F}$: acting at an angle 37° with the displacement vector. Given that $F = 100$ newton
   This force can be resolved into two components-
   (i) Horizontal component $F_x = 100 \cos 37°$
   (ii) Vertical component $F_y = 100 \sin 37°$
2. Gravitational force $M\vec{g}$: acting in vertically downward direction
   $$Mg = 10g = 100 \text{ newton}$$
3. Normal reaction $\vec{N}$: acting in vertically upward direction.
   $$N + 100 \sin 37° = Mg$$
   or $N = Mg - 100 \sin 37°$
   $$= 100 - 100\left(\frac{3}{5}\right) = 40 \text{ newton.}$$
4. Force of kinetic friction (f_k): acting opposite to displacement vector i.e., along negative direction of x axis.
   Force of kinetic friction,
   $$f_k = \mu_k N = (0.4)(40) = 16 \text{ newton}$$

As all above forces are constant, therefore, for each force use the work formula-
$$W = \vec{F}.\vec{s} = Fs\cos\theta$$

here, θ is the angle between force $\vec{F}$ and displacement $\vec{s}$ of point of application of the forec.

SOLUTION 1. Work done by applied pulling force:
$$W_1 = Fs \cos\theta = (100\ newton)(5m) \cos 37°$$
$$= 400\ J$$
2. Work done by the gravitational force:
$$W_2 = (Mg)s \cos 90°$$
$$= (100\ newton)(5\ m)(0) = 0\ J$$
3. Work done by the normal reaction:
$$W_3 = (N)s \cos 90°$$
$$= (40\ newton)(5\ m)(0) = 0\ J$$
4. Work done by force of kinetic friction:
$$W_4 = f_k s \cos\theta = (16\ newton)(5m) \cos 180°$$
$$= -80\ J$$

EXAMPLE 9. A particle is moving along a straight line from point A to point B with position vectors $2\hat{\imath} + 7\hat{\jmath} - 3\hat{k}\ m$ and $5\hat{\imath} - 3\hat{\jmath} - 6\hat{k}\ m$ respectively. One of the forces acting on the particle is $\vec{F} = 20\hat{\imath} - 30\hat{\jmath} + 15\hat{k}$ newton. Find the work done by this force.

APPROACH Given that, force $\vec{F} = 20\hat{\imath} - 30\hat{\jmath} + 15\hat{k}$ N
And displacement vector
$\vec{s} = (5\hat{\imath} - 3\hat{\jmath} - 6\hat{k}) - (2\hat{\imath} + 7\hat{\jmath} - 3\hat{k}) = 3\hat{\imath} - 10\hat{\jmath} - 3\hat{k}$
Here, force is a constant vector, therefore apply the formula for work, $W = \vec{F}.\vec{s}$

SOLUTION $W = \vec{F}.\vec{s}$
Substituting the given values in above equation, we get
$\therefore \quad W = (20\hat{\imath} - 30\hat{\jmath} + 15\hat{k}).(3\hat{\imath} - 10\hat{\jmath} - 3\hat{k})$
$= 60 + 300 - 45 = 315\ J$

EXAMPLE 10. Calculate work done by the force $\vec{F} = 3\hat{\imath} - 2\hat{\jmath} + 4\hat{k}\ N$, in carrying a particle from point $(-2m, 1m, 3m)$ to $(3m, 6m, -2m)$.

APPROACH Given that, force $\vec{F} = 3\hat{\imath} - 2\hat{\jmath} + 4\hat{k}\ N$
And displacement vector
$\vec{s} = (3 - (-2))\hat{\imath} + (6 - 1)\hat{\jmath} + (-2 - 3)\hat{k}$ meter
i.e., $\vec{s} = 5\hat{\imath} + 5\hat{\jmath} - 5\hat{k}$
Here, force is a constant vector, therefore apply the formula for work, $W = \vec{F}.\vec{s}$

SOLUTION $W = \vec{F}.\vec{s} = (3\hat{\imath} - 2\hat{\jmath} + 4\hat{k}).(5\hat{\imath} + 5\hat{\jmath} - 5\hat{k})$
$= 15 - 10 - 20 = -15J$

EXAMPLE 11. Two forces, $\vec{F}_1 = (2\hat{\imath} + 3\hat{\jmath} - \hat{k})N$ and $\vec{F}_2 = (\hat{\imath} - 2\hat{\jmath} + 2\hat{k})N$, are acting on a particle and shift it from point $(0, 0, 1\ m)$ to $(1\ m, 1\ m, 2\ m)$. Find the total work done by these forces.

APPROACH As the given forces are constant vector, therefore use the work formula, $W = \vec{F}_{net}.\vec{s}$

SOLUTION $\vec{F}_{net} = (2\hat{\imath} + 3\hat{\jmath} - \hat{k})N + (\hat{\imath} - 2\hat{\jmath} + 2\hat{k})N$
$= (3\hat{\imath} + \hat{\jmath} + \hat{k})\ N$
$\vec{s} = (1 - 0)\hat{\imath} + (1 - 0)\hat{\jmath} + (2 - 1)\hat{k}\ m$
$\vec{s} = (\hat{\imath} + \hat{\jmath} + \hat{k})\ m$
$\therefore W = (3\hat{\imath} + \hat{\jmath} + \hat{k}).(\hat{\imath} + \hat{\jmath} + \hat{k}) = 3 + 1 + 1 = 5\ J$

8. WORK ALONG A CURVED PATH

Figure 1 shows a curved path from P_1 to P_2. Suppose, a particle is confined to move along this curved path only. $\vec{F}$ is a force that acts on the particle. This force varies in direction as well as in magnitude.

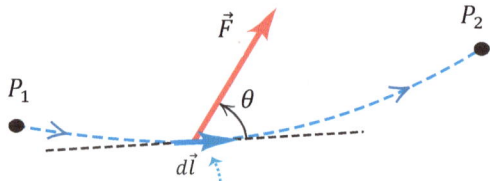

During an infinitesimal displacement $\vec{dl}$, the force $\vec{F}$ does work dW on the particle:
$$dW = \vec{F}.\vec{dl} = F \cos\theta\ dl$$
(a)

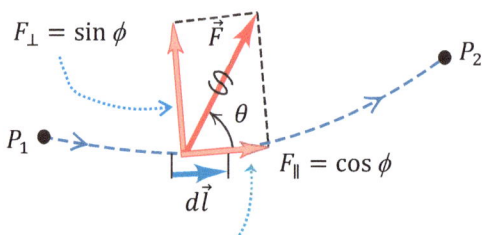

Only the component of $\vec{F}$ parallel to the displacement, $F_\parallel = F\cos\theta$, contributes to the work done by $\vec{F}$.
(b)

FIGURE. 1. A particle moves from point P_1 to P_2, acted on by a force $\vec{F}$ that varies in magnitude and direction.

To calculate the total work done by this force from point P_1 to P_2, we divide the curve between P_1 to P_2 into many infinitesimal vector displacements, and we call a typical one of these $\vec{dl}$. Each $\vec{dl}$ is tangent to the path at its position. Let $\vec{F}$ be the force at a typical point along the path, and let θ be the angle between $\vec{F}$ and $\vec{dl}$ at this point. Then, the small elementary work dW done by $\vec{F}$ on the particle during the displacement $\vec{dl}$:
$$dW = \vec{F}.\vec{dl} = F\cos\theta\ dl = F_\parallel dl$$
where $F_\parallel = F\cos\theta$ is the component of $\vec{F}$ in the direction parallel to $\vec{dl}$ (Fig. 1b).

Therefore, the total work done by $\vec{F}$ on the particle as it moves from P_1 to P_2 is
$$W = \int_{P_1}^{P_2} \vec{F}.\vec{dl} = \int_{P_1}^{P_2} F\cos\theta\ dl = \int_{P_1}^{P_2} F_\parallel dl$$

☞ *Note that only the component of the net force parallel to the path, $F_\parallel$, does work on the particle, so only this component can change the speed and kinetic energy of the particle. The component perpendicular to the path, $F_\perp = F\sin\theta$, has no effect on the particle's speed; it acts only to change the particle's direction.*

EXAMPLE 12. A block of mass m is taken from A to B along spherical smooth bowl of radius R. Calculate the work done by each force acting on the block.

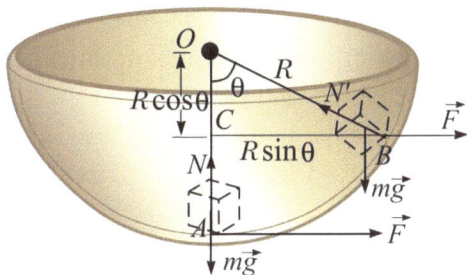

APPROACH To calculate the work done corresponding to any force on the block, find the displacement in the direction of each force and multiply it by corresponding force.
SOLUTION
1. Forces on the block-
(i) $m\vec{g}(\downarrow)$ (Gravitational force on the block in downward direction)
(ii) $\vec{F}(\rightarrow)$ (Horizontally applied force on the block along right side)
(iii) $\vec{N}$ (Normal reaction of the surface towards origin at each point of the surface)

2. Displacements-
(i) Displacement of block along vertical direction $AC = R(1 - \cos\theta) \uparrow$ (upwards)
(ii) Displacement of block along horizontal direction $CB = R \sin\theta \;(\rightarrow)$
Work Done by gravity:
Since, the vertical displacement and force of gravity $(m\vec{g})$ are oppositely directed, therefore,
(i) Work done by force of gravity $= -mg R(1 - \cos\theta)$
(ii) Work Done by force $F = F(R \sin\theta) = FR\sin\theta$
Work Done by normal reaction $= 0$
($\because$ the normal reaction $\vec{N}$, is continuously perpendicular to the tangential displacement)

EXAMPLE 13. A small particle of mass m is pulled to the top of a frictionless half cylinder (of radius R) by a light cord that passes over the top of the cylinder as illustrated in Fig. 1. (a) Assuming the particle moves at a constant speed, show that $F = mg \cos\theta$. Note: If the particle moves at constant speed, the component of its acceleration tangent to the cylinder must be zero at all times. (b) By directly integrating $W = \int \vec{F}.d\vec{l}$, find the work done by F, in moving the particle at constant speed from the bottom to the top of the half-cylinder.

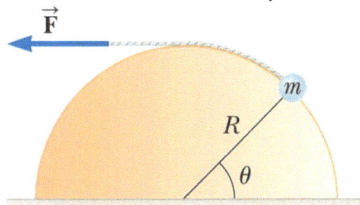

FIGURE 1
APPROACH As the tension in a string always acts along its length, therefore, the pulling force $\vec{F}$, will always be tangential to the cylindrical curved path.

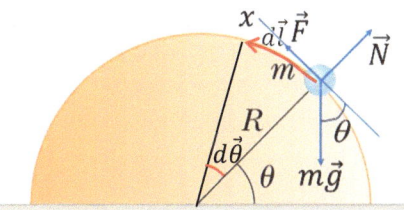

FIGURE. 2
Now, if the particle moves with constant speed, then net tangential force acting on the particle must be zero, i.e., $\Sigma F = 0$. In this case, the net work done on the particle would also be zero.
If dl, is small tangential displacement of the particle, then small work done by force $\vec{F}$, $dW = F_\parallel dl$.
If $d\theta$ is small angular displacement, corresponding to tangential displacement dl [see Fig. 2], then $dl = R\,d\theta$.
Therefore, $dW = F_\parallel (R\,d\theta)$
Therefore, the net work done, corresponding to angular displacement $\theta = 0$ to top most point, i.e., $\theta = \frac{\pi}{2}$ will be given by
$$W = \int_0^{\pi/2} F_\parallel (R\,d\theta)$$
For given diagram, $F_\parallel = F$, therefore, $W = \int_0^{\pi/2} F(R\,d\theta)$
SOLUTION (a) The radius to the object makes angle θ with the horizontal. Taking the x axis in the direction of motion tangent to the cylinder, the object's weight makes an angle θ with the $-x$ axis. Then,
$$\Sigma F_x = 0$$
$\Rightarrow \quad F - mg\cos\theta = 0$
or $\quad F = mg\cos\theta$
(b) $W = \int_0^{\pi/2} F(R\,d\theta)$
Here, $F = mg\cos\theta$
$\therefore \qquad W = \int_0^{\pi/2} mg\cos\theta\, R\,d\theta$
$\qquad\qquad = mgR\,[\sin\theta]_0^{\pi/2} = mgR(1 - 0)$
$\qquad\qquad = mgR$

☞ In this case, the work done by gravity,

$W_{grav} = -\int_0^{\pi/2} mg\cos\theta\, R\,d\theta$
Here $-ve$ sign indicates that, the tangential force applied by gravity, $mg\cos\theta$, is opposite to the elementary displacement vector $d\vec{l}$
Solving above integral, we get
$$W_{grav} = -mgR\,[\sin\theta]_0^{\pi/2} = -mgR(1 - 0)$$
$\qquad\qquad = -mgR$
Therefore, the net work done on the particle, $W = W_F + W_{grav} = mgR + (-mgR) = 0$.

9. CHECKPOINT 1

1. •A 3000 kg truck is to be loaded onto a ship by a crane that exerts an upward force of $31 kN$ on the truck. This force, which is strong enough to overcome the gravitational force and keep the truck moving upward, is applied over a distance of 2.0 m.

MECHANICS

Find (a) the work done on the truck by the crane, (b) the work done on the truck by gravity, and (c) the net work done on the truck.

2. A block kept on rough surface is being pulled by force $\vec{F}$, as shown in following figure. Find whether the work done by –

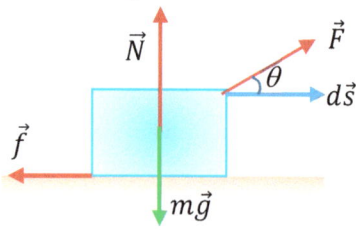

(i) pulling force $\vec{F}$.
(ii) friction.
(iii) Normal reaction and weight
are positive, negative or zero.

3. •A student holds her 1.5-kg physics textbook out a second-story dormitory window until her arm is tired; then she releases it. (a) How much work is done on the book by the student in simply holding it out the window? (b) How much work is done by the force of gravity during the time in which the book falls $3.0\ m$?

4. ••A worker pulls am m-kg crate with a rope, as illustrated in following figure. The coefficient of kinetic (sliding) friction between the crate and the floor is μ_k. If he moves the crate with a constant velocity a distance of s, how much work is done?

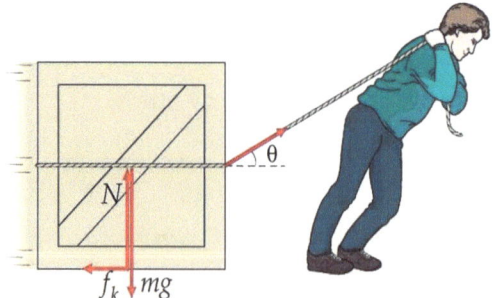

5. ••In above problem, if mass of the crate is 40.0-kg, the coefficient of kinetic (sliding) friction between the crate and the floor is 0.55. and the distance is $7.0\ m$, then calculate the work done in displacing it by a distance $7\ m$?

6. •A cyclist comes to a skidding stop in $10\ m$. During this process, the force on the cycle due to the road is $200\ N$ and is directly opposed to the motion. (a) How much work does the road do on the cycle? (b) How much work does the cycle do on the road?

7. •A force $(3\hat{\imath} + 4\hat{\jmath})$ newton acts on a body and displaced it by $(3\hat{\imath} + 4\hat{\jmath})$ metre. What is the work done by this force?

8. ••Figure shows three forces applied to a trunk that moves leftward by $3m$ over a smooth floor. The force magnitudes are $F_1 = 5N$, $F_2 = 9N$, and $F_3 = 3N$. Find the net work done on the trunk by the three forces.

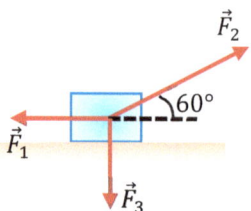

9. ••A block of mass m is pushed up a rough inclined plane of angle θ by a constant force $\vec{F}$ parallel to the incline, as shown in following figure. The displacement of the block up the incline is $\vec{s}$

(a) Find the work done by: the force $\vec{F}$, the kinetic friction $\vec{f}_k$, the force of gravity $m\vec{g}$, and the normal force $\vec{N}$.

(b) Calculate the work done of part (a) for $m = 2\ kg$, $\mu_k = 0.5$, $\theta = 30°$, $F = 20\ N$, and $s = 5\ m$.

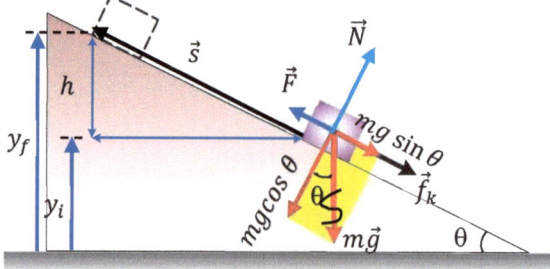

10. ••A $10\ kg$ block is kept on an inclined plane, of inclination $30°$, which is placed inside a lift moving upward with constant velocity $2\ m/s$. The block is at rest w.r.t. the inclined plane. Calculate the work done on the block by frictional force, weight of the block and normal reaction in $5\ s$.

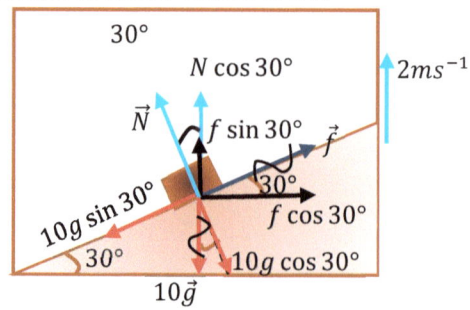

11. •A body is displaced from $(0,0,0)$ to $(3,-2,4)m$ using the force $(4\hat{i}+3\hat{j}-\hat{k})N$. Calculate the work done by this force.

12. ••A block is at rest with respect to a cart. Calculate the work done by the friction in $2s$. Initial velocity = 0

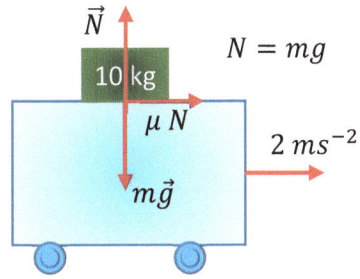

10. WORK DONE BY A VARIABLE FORCE

So far, we have dealt with the work done by constant forces. However, most of the forces that we encounter in daily life are not constant. Important examples include electric, gravitational and spring forces which vary with the distance between interacting objects.

So, it is always possible to have forces whose magnitude or direction or both vary from point to point. In such cases, a generalization of work equation is needed. One dimensional and three-dimensional generalizations are given below-

10.1. ONE-DIMENSIONAL ANALYSIS

Consider an object that is being displaced along the x-axis from x_i to x_f due to the application of a varying positive force $F(x)$, as shown in Fig. 1a. To calculate the work done by this force, we imagine that the object undergoes a very small displacement Δx from x to $x + \Delta x$ due to the effect of an approximate constant force $F(x)$ as shown in Fig.1b. For this very small displacement, we represent the amount of work done by the force by the expression:

$$\Delta W = F(x)\Delta x \qquad \ldots (1)$$

which is just the area of the magnified rectangle shown in Fig.1b. Then, the total work done from x_i to x_f by the variable force $F(x)$ is approximately equal to the sum of the large number of rectangles in Fig.2b, i.e. the total area under the force curve. Thus:

$$W = \sum_{x_i}^{x_f} F(x)\,\Delta x \qquad \ldots (2)$$

In the limit where Δx approaches zero, the value of the sum in the last equation approaches the exact value of the area under the force curve, see Fig.1c. As you probably know from calculus, the limit of that sum is called an integral and is represented by:

$$\lim_{\Delta x \to 0} \sum_{x_i}^{x_f} F(x)\,\Delta x = \int_{x_i}^{x_f} F(x)\,dx \qquad \ldots (3)$$

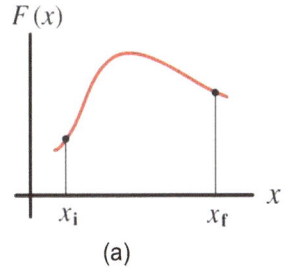

(a)

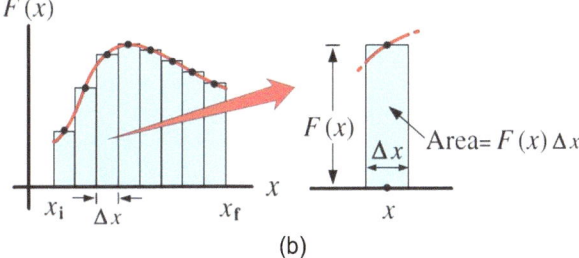

(b)

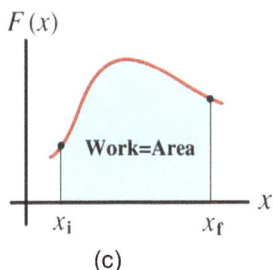

(c)

FIGURE. 1 (a) A variable force F(x) displaces a body in the positive x direction from x_i to x_f. (b) The area under the curve is divided into narrow strips of thickness Δx, so that the approximate work done by the force F(x) for the small displacement Δx is $\Delta W = F(x)\Delta x$. (c) In the limiting case, the work done by the force is the shaded area under the force curve

Therefore, we can express the work done by a variable force $F(x)$ on an object that undergoes a displacement from x_i to x_f as follows:

$$W = \int_{x_i}^{x_f} F(x)\,dx \qquad \ldots (4)$$

If $F(x)$ is positive in some regions and negative in others, then the final sum of all areas with signs is called the net signed area and is equal to the area of the regions where $F(x) > 0$ minus the area of the regions where $F(x) < 0$.

10.2. THREE-DIMENSIONAL ANALYSIS

Consider a particle that is acted upon by a three-dimensional force of the following form:

$$\vec{F} = F_x\hat{i} + F_y\hat{j} + F_z\hat{k} \qquad \ldots (1)$$

where the components F_x, F_y, and F_z are generally a function of the position vector of the particle. Furthermore, let the particle move from position $\vec{r} = x\hat{i} + y\hat{j} + z\hat{k}$, to $\vec{r} + d\vec{r} = (x+dx)\hat{i} + (y+dy)\hat{j} + (z+dz)\hat{k}$, through an incremental displacement $d\vec{r}$, i.e.

$$d\vec{r} = dx\hat{i} + dy\hat{j} + dz\hat{k} \qquad \ldots (2)$$

In this case, the increment of work dW done on the particle by the force $\vec{F}$ during the incremental displacement $d\vec{r}$ is giving by:

$$dW = (F_x\hat{i} + F_y\hat{j} + F_z\hat{k})\cdot(dx\hat{i} + dy\hat{j} + dz\hat{k})$$
$$= F_x dx + F_y dy + F_z dz \qquad \ldots (3)$$

The work W done by the force $\vec{F}$ on the particle when it moves from an initial position $\vec{r_i}$ of coordinates (x_i, y_i, z_i) to a final position $\vec{r_f}$ of coordinates (x_f, y_f, z_f) can be represented by:

$$W = \int_{r_i}^{r_f} dW = \int_{x_i}^{x_f} F_x dx + \int_{y_i}^{y_f} F_y dy + \int_{z_i}^{z_f} F_z dz \qquad \ldots (4)$$

When $\vec{F}$ has only an x component, this equation reduces to $W = \int_{x_i}^{x_f} F_x dx$

EXAMPLE 14. A force which varies with position coordinate x according to equation $F_x = (4x + 2)$ N. Here x is in meters. Calculate work done by this force in carrying a particle from position $x_i = 1\,m$ to $x_f = 2\,m$.

APPROACH 1. Since the given force is a variable force, therefore, use $W = \int_{x_i}^{x_f} F_x dx$ to calculate work done by force F_x during displacement of point of application of force from x_i to x_f.

SOLUTION Using the equation $W = \int_{x_i}^{x_f} F_x dx$

$$W = \int_{1m}^{2m} (4x + 2)\,dx$$
$$= [2x^2 + 2x]_{1m}^{2m}$$
$$= [2(2^2) + 2(2)] - [2(1^2) + 2(1)]$$
$$= 12 - 4 = 8\,J$$

APPROACH 2. Graphical approach: we can also solve the above problem by graphical method.
If A is the area of the F_x vs x graph on x axis, from $x = x_i$ to $x = x_f$, then, the work done by force F_x-
$$W = A$$
So first draw the graph $F_x = (4x + 2)$, then calculate its area on position axis. This signed area will be the required work on the particle.

SOLUTION Area covered by graph on x axis, from $x = 1m$ to $x = 2m$,
$$W = \text{shaded trapezium area on the } x \text{ axis}.$$

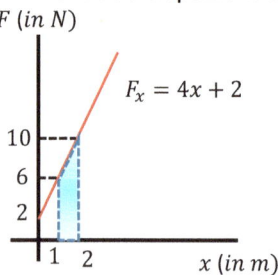

i.e., $W = \frac{1}{2}(6 + 10) \times 1 = 8\,J$
Therefore, the work done by force F_x,
$$W = 8\,J$$

EXAMPLE 15. A horizontal force F is used to pull a box placed on floor. Variation in the force with position coordinate x measured along the floor is shown in the graph.

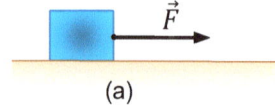
(a)

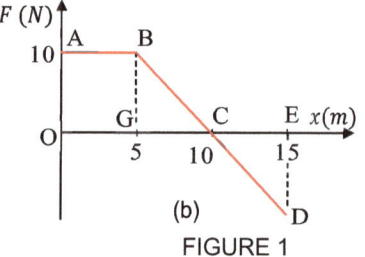
(b)

FIGURE 1

(a) Calculate work done by the force in moving the box from $x = 0\,m$ to $x = 10\,m$.
(b) Calculate work done by the force in moving the box from $x = 10\,m$ to $x = 15\,m$.
(c) Calculate work done by the force in moving the box from $x = 0\,m$ to $x = 15\,m$.

APPROACH Since, in a rectilinear motion, the work done by a force is always equal to area under the force-position graph and the position axis, therefore to find work done, calculate the required area by using geometry. It is to be noted that, the area below position axis is considered as negative.

SOLUTION
(a) $W_{0 \to 10\,m}$ = Area of trapezium OABC
$$= \frac{1}{2}(AB + OC) \times OA = \frac{1}{2}(5 + 10) \times 10 = 75\,J$$
(b) $W_{10m \to 15}$ = $-$ Area of triangle CDE
$$= \frac{1}{2} CE \times ED = \frac{1}{2} \times 5 \times 10 = -25\,J$$
(c) $W_{0 \to 15}$ = $-$ Area of trapezium OABC $-$ Area of triangle CDE $= 75\,J - 25\,J = 50\,J$

EXAMPLE 16. Calculate work done by the force $\vec{F} = x\hat{i} + y^2\hat{j}$ N, in carrying a particle from point $(1m, 2m)$ to $(-3m, 4m)$.

APPROACH In this problem, the force is variable in nature, therefore use the formula, $dW = \vec{F}\cdot d\vec{s}$, with $\vec{F} = x\hat{i} + y^2\hat{j}$ and $d\vec{s} = dx\hat{i} + dy\hat{j}$. Then integrate it from, $(1, 2)$ to $(-3, 4)$.

SOLUTION $dW = \vec{F}\cdot d\vec{s}(d\vec{s} = dx\hat{i} + dy\hat{j})$
$dW = x\,dx + y^2\,dy$
$$W = \int_1^{-3} x\,dx + \int_2^4 y^2\,dy = \frac{x^2}{2}\Big|_1^{-3} + \frac{y^3}{3}\Big|_2^4$$
$$= \left[\frac{9}{2} - \frac{1}{2}\right] + \left[\frac{64}{3} - \frac{8}{3}\right] = 4 + \frac{56}{3} = \frac{68}{3}J = 22.67\,J$$

EXAMPLE 17. A particle is moving on circle of radius $5\,m$. A force of constant magnitude of $10\,N$ is acting on particle along the tangent. Find the work done by this force when particle completes one cycle.

MISCONCEPTION Work is zero because displacement is zero.

APPROACH Force $\vec{F}$ is varying because its direction is changing so $W \neq \vec{F}\cdot\vec{s}$
Calculate the work done by tangential force by using the expression
$$dW = \vec{F}\cdot d\vec{s} = Fds$$
where, ds is a small path length and small path length will always be along tangent

$$\Rightarrow W = \int F ds$$

SOLUTION Work done by tangent force
$$W = F \int ds = F \times 2\pi r$$
$$= 10 \times 2 \times \pi \times 5 = 100\pi J$$

EXAMPLE 18. A block of mass $10\, kg$ is pulled by a force $\vec{F}$ having magnitude $20N$. Find work done by force F if body moves $5m$ in right direction. Given that $a = 25\, m$ and $b = 1m$ initially.

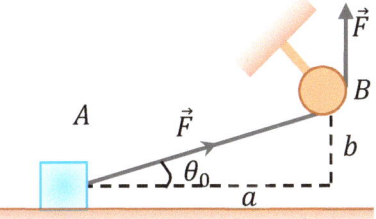

APPROACH In this case, the force $\vec{F}$ depends on angle θ with x axis, therefore apply the concept of variable force with the proper selection of displacement.

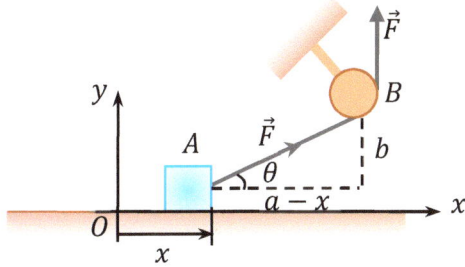

SOLUTION Let us consider the positive direction of x axis towards right and positive y direction upwards. Suppose at any instant t, the distance travelled by the block is s. If at this time, the force vector $\vec{F}$ makes angle θ with the positive direction of x axis, then from adjoining figure, we have
$$\vec{F} = F\cos\theta\,\hat{\imath} + F\sin\theta\,\hat{\jmath} \quad \ldots (1)$$
Now, if just after time t, we consider a small time interval dt and in that time interval the block moves a distance dx, then from the adjoining figure, the small work done by force $\vec{F}$
$$dW = \vec{F}.d\vec{x} = (F\cos\theta\,\hat{\imath} + F\sin\theta\,\hat{\jmath}).dx\,\hat{\imath}$$
or $\qquad dW = (F\cos\theta)dx \quad \ldots (2)$

Again, from adjoining figure, $\cos\theta = \dfrac{a-x}{\sqrt{(a-x)^2+b^2}}$

Substituting this value of $\cos\theta$, in Eq. (2), we get
$$\therefore\quad dW = F\frac{a-x}{\sqrt{(a-x)^2+b^2}}dx$$

Therefore, total work done by force $\vec{F}$, in displacing the block by $5\, m$.
$$W = \int_0^5 F\frac{a-x}{\sqrt{(a-x)^2+b^2}}dx \quad \ldots (3)$$

Let, $(a-x)^2 + b^2 = t$
$\therefore\quad -2(a-x)dx = dt$
or $\quad (a-x)dx = -\dfrac{1}{2}dt$

and when $x = 0$, $t = a^2 + b^2$
when, $x = 5$, $t = (a-5)^2 + b^2$

Substituting these values in equation (3), we get

$$W = -\frac{1}{2}F\int_{a^2+b^2}^{(a-5)^2+b^2}\frac{dt}{\sqrt{t}}$$

or $\quad W = -\dfrac{1}{2}F\left[\dfrac{\sqrt{t}}{1/2}\right]_{a^2+b^2}^{(a-5)^2+b^2}$

or $\quad W = -F[\sqrt{(a-5)^2+b^2} - \sqrt{a^2+b^2}]$

or $\quad W = F[\sqrt{a^2+b^2} - \sqrt{(a-5)^2+b^2}]$

Substituting the given values of a, b and F, we get
$$W = 20\left[\sqrt{25^2+1^2} - \sqrt{(25-5)^2+1^2}\right]J$$

or $\quad W = 20[\sqrt{624} - \sqrt{401}]J$

10.3. WORK DONE BY SPRING FORCE

Let us consider a spring block system as shown in Fig. 1. Suppose the natural length of the spring is l_0. If the spring is stretched or compressed, the spring exerts a force on the block.

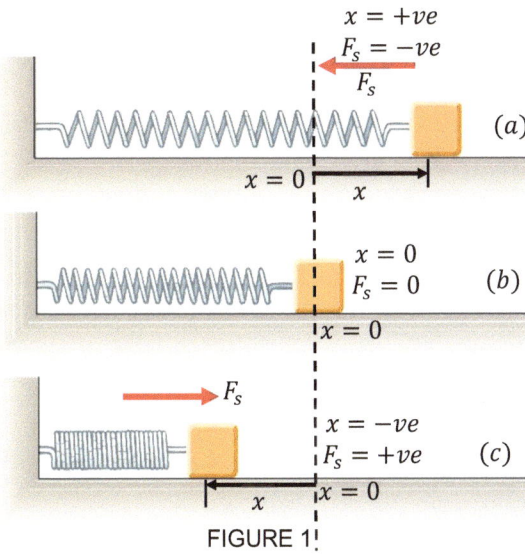

FIGURE 1

For displacement x, from the natural length, the spring force is given by
$$F_s = -kx \quad \text{(Hooke's Law)} \quad \ldots (1)$$
Here, k is a constant characteristic of the particular spring, known as the spring constant.

Note the sign: The spring force always acts to bring the mass back to $x = 0$. When x is positive (the spring is stretched toward the right on the x axis), the mass is on the right side in Fig. 1(a) and the force F_s is negative (it is a pull toward the left), and when x is negative (the spring is compressed toward the left), the mass is on the left side of the origin (Fig. 1(c)) and the force F_s is positive (it is a push toward the right).

Note that a spring force is a variable force because it is a function of x, the position of the free end. Thus, F_s can also be written as $F(x)$.

Also note that Hooke's law is a linear relationship between F_s and x [Fig.2].

14 MECHANICS

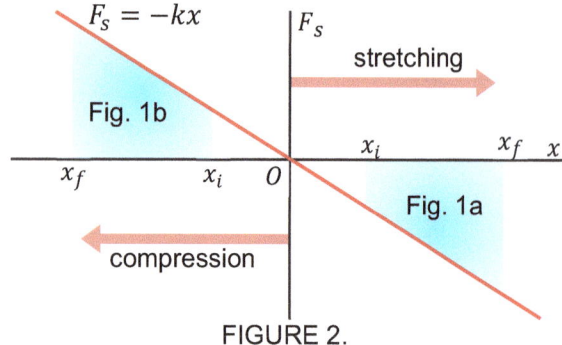

FIGURE 2.

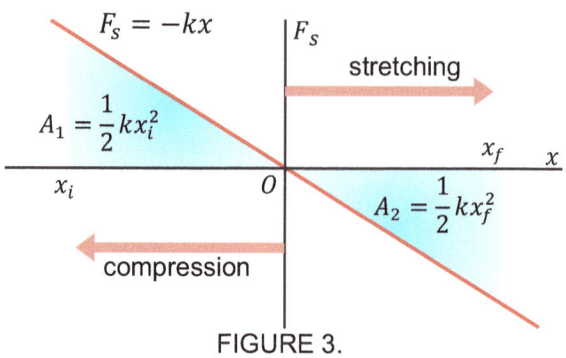

FIGURE 3.

If at any instant the expansion in spring is x, then spring force on the block in vector form
$$\vec{F_s} = -kx\hat{\imath} \qquad \ldots (2)$$
[As the spring force is acting against expansion, i.e., in $= -ve$ direction of x axis]
If we further pull the block by infinitesimal displacement dx, then small work done by spring,
$$dW_s = \vec{F_s}.\vec{dx}.$$
or $\quad dW_s = (-kx\hat{\imath}).(dx\hat{\imath}) = -kx\,dx$
∴ the work done by spring force in compression from $x = x_i$ to $x = x_f$ is given by
$$W_s = -\int_{x_i}^{x_f} kx\,dx = -k\left[\frac{x^2}{2}\right]_{x_i}^{x_f}$$

or $\quad W_s = -\frac{1}{2}k[x_f^2 - x_i^2]$

or $\quad W_s = \frac{1}{2}k[x_i^2 - x_f^2] \qquad \ldots (3)$

From Eq. (1), it is clear that-
$$W_s = \begin{cases} 0, & \text{if } x_f = x_i \\ > 0, & \text{if } x_f < x_i \\ < 0, & \text{if } x_f > x_i \end{cases} \qquad \ldots (4)$$

i.e. work W_s is positive if the block ends up closer to the relaxed position ($x = 0$) than it was initially. It is negative if the block ends up farther away from $x = 0$. It is zero if the block ends up at the same distance from $x = 0$.

In other words, the work W_s done by the spring force can have a positive or negative value, depending on whether the net transfer of energy is to or from the block as the block moves from x_i to x_f.

The magnitudes of shaded areas in Fig. 2 represent the negative work done by the spring in Figs. 1a and 1b.
Note: If initially, the spring was compressed (i.e., x_i is negative) and finally it is stretched (i.e., x_f is positive), then, from Fig.3, the work done by spring will be given by-
$$W_s = |A_1| - |A_2| = \frac{1}{2}kx_i^2 - \frac{1}{2}kx_f^2$$

Now, suppose you pull on an initially relaxed spring (i.e., $x_i = 0$), stretching it to a final extension x (i.e., $x_f = x$). Then, from equation (3), the work done by spring force will be given by
$$W_s = -\frac{1}{2}kx^2 \quad \text{(work by a spring force)}$$

EXAMPLE 19. A spring of stiffness k, was initially compressed by x_1. If it is further compressed by x_2, then find the work done by spring.
APPROACH The work done by spring force is given by
$$W_s = \frac{1}{2}k[x_i^2 - x_f^2] \qquad \ldots (1)$$
Since, the spring was initially compressed by x_1 and further compressed by x_2
∴ $x_i = x_1$, $x_f = x_1 + x_2$
Now, substitute these values in (1) and solve for W_s.
SOLUTION $W_s = \frac{1}{2}k[x_1^2 - (x_1 + x_2)^2]$
or $\quad W_s = -\frac{1}{2}k[x_2^2 + 2x_1x_2]$

11. CHECKPOINT 2

1. •A force F_x acts on a particle. The force is related to the position of the particle by the formula $F_x = cx^3$, where c is a constant. Find the work done by this force on the particle when the particle moves from $x = 1.5\,m$ to $x = 3\,m$.
   ☞ The work done by this force as it displaces the particle is the area under the curve of F as a function of x. Note that the constant c has units of N/m^3.

2. •A can of sardines is made to move along an x axis from $x = 0.25\,m$ to $x = 1.25\,m$ by a force with a magnitude given by $F = e^{-4x^2}$, with x in meters and F in newtons. How much work is done on the can by the force?

3. ••A person decided to carry some water up a $20\,m$ high tower. His bucket has a mass of $10\,kg$ and holds $30\,kg$ of water when it is full. However, the bucket has a hole, and as the person climbed at a constant speed, water leaked out at a constant rate. When he got to the top, only $10\,kg$ of water

remained in the bucket. (a) Write an expression for the mass of the bucket plus water as a function of the height (y) climbed. (b) Find the work done by person on the bucket.

12. DEPENDENCE OF WORK ON REFERENCE FRAME

A force is always independent of reference frame and remains same in all frames of references, but displacement depends on frame of reference and its value will be different with respect to different frame of references. Since, work depends on force and displacement both, therefore, the work of a force also depends on choice of reference frame. For example, consider a porter, with a suitcase on his head, moving up a staircase up to a vertical height h. The work done by the upward lifting force on the suitcase, relative to him, will be zero (as displacement relative to him is zero). while relative to a person on the ground will be $\vec{F}.\vec{s} = mgh$ (as $F = mg$ and $s = h$)
which is a nonzero quantity.

13. WORK-ENERGY THEOREM AND DEFINITION OF KINETIC ENERGY

Let us consider an inertial reference frame. Suppose, in this inertial frame, a particle of mass m is moving with acceleration $a = a(x)$ along the x-axis under the effect of a net force $F(x)$ that points along this axis. Thus, according to Newton's second law of motion we have $F(x) = ma$. If we denote, the initial position by i and final position by f, then the work done by this net force on the particle as it moves from an initial position x_i to a final position x_f can be found as follows:

$$W = \int_{x_i}^{x_f} F(x)dx = \int_{x_i}^{x_f} ma\, dx \qquad \ldots (1)$$

The acceleration a, as a function of x can be written as
$$a = v.\frac{dv}{dx}$$

Substituting this result back into Eq.1 yields:
$$W = \int_{x_i}^{x_f} ma\, dx = \int_{x_i}^{x_f} m\left(v.\frac{dv}{dx}\right) dx$$

If v_i and v_f are the velocities of the particle at initial and final positions respectively, then

or $\qquad W = \int_{v_i}^{v_f} mv\, dv = m \int_{v_i}^{v_f} v\, dv$

or $\qquad W = m\left[\frac{1}{2}v^2\right]_{v_i}^{v_f} = \frac{1}{2}mv_f^2 - \frac{1}{2}mv_i^2 \qquad \ldots (2)$

Thus, the total work done is equal to the change in the quantity $\frac{1}{2}mv^2$, which is called the object's **translational kinetic energy** (symbol K). (Often, we just say *kinetic energy* if it is understood that we mean translational kinetic energy.) Translational kinetic energy is the energy associated with motion of the object as a whole; it does not include the energy of rotational or internal motion. Therefore, we can write-

Translational Kinetic Energy

The kinetic energy K of a particle is defined as the product of one half of its mass and the square of its speed, i.e.,

$$K = \frac{1}{2}mv^2 \qquad \ldots (3)$$

(here $v \ll c$, the speed of light)

Kinetic energy is a scalar quantity and has the same units as work. In SI units we have:

$1J = 1 kg.m^2/s^2 = 1 N.m$

Again, $K = \frac{1}{2}mv^2 = \frac{1}{2m}(mv)^2 = \frac{p^2}{2m}$

here, $p = mv$ (linear momentum of the particle). Therefore, the kinetic energy of a particle in terms of linear momentum can be given as

$$K = \frac{p^2}{2m} \text{ (KE in terms of linear momentum)} \qquad \ldots (4)$$

We can view kinetic energy as the energy associated with the motion of an object. It is more convenient to express eq. (2) as:

$$W = \Delta K = K_f - K_i \qquad \ldots (5)$$

where K_i is the particle's initial kinetic energy and K_f is its final kinetic energy after the work is done. Thus, the work done by the *net force* in displacing a particle equals the change in its kinetic energy.

Note that W, in Eq. (5), is net work done by all forces not just a particular force acting on the particle.

If there are many forces, such as an applied force $\vec{F}$, a gravitational force F_g, a spring force $\vec{F}_s$ a frictional force f, etc., then, the work done by the net force in displacing a particle will be equal to the sum of the work done by all the forces acting on the particle. That is:

$W_{net} = W_F + W_g + W_s + W_f + \ldots = K_f - K_i = \Delta K$

or $\qquad W_{net} = K_f - K_i = \Delta K \qquad \ldots (6)$

(Work-Energy theorem)

☞ Equation (6) is known as the **work-energy theorem**. *This theorem is valid in all inertial frames even when the force varies in direction and magnitude while the particle (or the object) moves along an arbitrary curved path in three dimensions.*

Kinetic energy is a scalar quantity and is always positive if the object is moving or zero if it is at rest. Kinetic energy can never be negative, although a *change* in kinetic energy can be negative. *The kinetic energy of an object moving with speed v is equal to the work that must be done on the object to accelerate it to* that speed starting from rest. When the total work done is positive, the object's speed increases, increasing the kinetic energy. When the total work done is negative, the object's speed decreases, decreasing the kinetic energy.

13.1. HOW TO APPLY WORK ENERGY THEOREM

Since, we have deduced the work energy principle for a particle moving relative to an inertial frame, therefore, it is advised to apply the work–energy theorem only to bodies that we can represent as particles—that is, as moving point masses, in an inertial reference frame.

WORKING STEPS: To use work energy theorem the following steps should be followed.
➢ Identify the initial and final positions clearly and write expressions for kinetic energies at these positions, whether known or unknown.
➢ Draw the FBD of the body at any intermediate stage between the initial and final positions. The forces shown will help in deciding their work. Calculate the work done by each force and add them to obtain net work done W by all the forces.
➢ Now, substitute the values of net work, initial and final kinetic energies in the expression.
$$W_{net} = K_f - K_i$$

EXAMPLE 20. A box of mass $m = 10$ kg is initially at rest on a rough horizontal surface, where the coefficient of kinetic friction between the box and the surface is $\mu_k = 0.2$. The box is then pulled horizontally by a force $F = 50N$ that makes an angle $\theta = 60°$ with the horizontal, see Fig.1. (a) Use the work-energy theorem to find the speed v_f of the box after it moves a distance of $4m$. (b) Repeat part (a) using Newtonian mechanics.

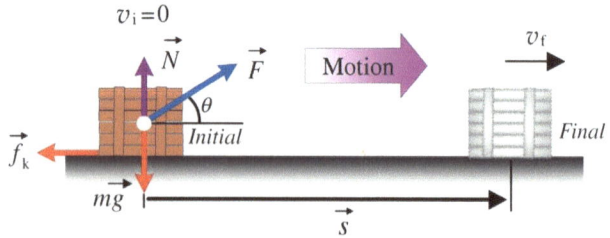

APPROACH (a) To find the final speed of the box, calculate net work done on the box, then apply the work energy theorem. Simplify it for final speed of the box.
(b) To find the final speed by using Newtonian mechanics, first find net acceleration a of the box with the help of $\Sigma F_x = ma$, then use the kinematic equation $v_f^2 = v_i^2 + 2as$.
SOLUTION (a) Both the weight-gravitational force $m\vec{g}$ and the normal force $\vec{N}$ do no work, since the displacement is horizontal, i.e. $W_g = W_N = 0$. The work done by the applied force is:
$W_F = \vec{F} \cdot \vec{s} = Fs \cos\theta = (50N)(4m)(\cos 60°) = 100J$
The magnitude of the frictional force is $f_k = \mu_k N$, where in this case $N = mg - F\sin\theta$. Therefore, the work done by friction is:
$W_f = \vec{f_k} \cdot \vec{s} = f_k s \cos 180° = \mu_k(mg - F\sin\theta)s(-1)$
$= -0.2 \times [(10\,kg)(9.8\,m/s^2) - (50N)(0.866)](4m)$
$= -43.76J$
Thus, the net work done on the box is:
$W_{net} = W_F + W_g + W_N + W_f$
$= 100J + 0 + 0 + (-43.76J)$
$= 56.24J$
Applying the work-energy theorem with $v_i = 0$ gives:
$W_{net} = K_f - K_i = \frac{1}{2}mv_f^2$
$\Rightarrow v_f = \sqrt{\frac{2W_{net}}{m}} = \sqrt{\frac{2 \times 56.24\,J}{10kg}} = 3.35\,m/s$
(b) Applying Newton's second law in the component form, then for the horizontal component, we find that:
$\Sigma F_x = F\cos\theta - f_k = ma$
Thus, the acceleration of the box will be given by:
$a = \frac{F\cos\theta - \mu_k(mg - F\sin\theta)}{m}$
$= \frac{(50N)(\cos 60°) - (0.2)[(10kg)(9.8m/s^2) - (50N)(0.866)]}{10\,kg}$
$= 1.406\,m/s^2$
To find the final speed, we use the kinematic equation $v_f^2 = v_i^2 + 2as$, when $v_i = 0$ to get:
$v_f = \sqrt{2as} = \sqrt{2 \times (1.406\,m/s^2)(4m)}$
$= 3.35\,m/s$
Because the forces are constants in this example, the analysis used by Newtonian mechanics is easier than that of the work-energy theorem.

EXAMPLE 21. A particle of mass $10\,kg$ is moving with velocity $2\hat{\imath} + 3\hat{\jmath} + 4\hat{k}$ under a constant force $\vec{F} = \hat{\imath} - \hat{\jmath} - 2\hat{k}$. Find the work done just after 2 second of start of the motion. Also calculate KE and velocity of the particle, just after 2 second of start of the motion.
APPROACH Calculating Work: The work done can be calculated by using the relation,
$$W = \vec{F} \cdot \vec{s} \quad \ldots (1)$$
Here, force $\vec{F}$ is known but displacement $\vec{s}$ is unknown. You can find $\vec{s}$ by applying kinematic equation, $\vec{s} = \vec{v_0}t + \frac{1}{2}\vec{a}t^2$.

Calculating Kinetic Energy and Velocity after 2 seconds
As, initial velocity is known, therefore, final velocity can be calculated by either applying work energy theorem
$$W = K_f - K_i \quad \ldots (2)$$
or by using kinematic equation,
$$\vec{v} = \vec{v_0} + \vec{a}t \quad \ldots (3)$$

SOLUTION Acceleration, $\vec{a} = \frac{\vec{F}}{mass} = \frac{i-j-2k}{10} m/s^2$

$\vec{s} = \vec{v}_0 t + \frac{1}{2}\vec{a}t^2 = (2\hat{i} + 3\hat{j} + 4\hat{k}) \times 2$

$\qquad + \frac{1}{2} \times \left(\frac{\hat{i}-\hat{j}-2\hat{k}}{10}\right) \times 4 = \frac{21}{5}\hat{i} + \frac{29}{5}\hat{j} + \frac{38}{5}\hat{k}$

$\therefore$ Work $W = \vec{F}.\vec{s} = (\hat{i} - \hat{j} - 2\hat{k}).\left(\frac{21}{5}\hat{i} + \frac{29}{5}\hat{j} + \frac{38}{5}\hat{k}\right)$

$= \frac{21}{5} - \frac{29}{5} - \frac{76}{5} = -\frac{84}{5} J$

Calculation of KE after 2 Second

Given that, initial velocity of the particle,

$\vec{v}_0 = 2\hat{i} + 3\hat{j} + 4\hat{k}$

$\therefore \qquad |\vec{v}_0| = \sqrt{4 + 9 + 16} = \sqrt{29}$

Now, by work energy theorem, we have

$W = K_f - K_i = K_f - \frac{1}{2}mv_0^2$

i.e., $-\frac{84}{5} = K_f - \frac{1}{2} \times 10 \times 29$

or $K_f = -\frac{84}{5} + 145 = \frac{641}{5}$

Calculation of Velocity After 2 Seconds.

If $\vec{v}$ is the velocity of the particle, just after 2 second, then-

$K_f = \frac{1}{2}mv^2$

or $\quad v = \sqrt{\frac{2K_f}{m}} = \sqrt{\frac{2\left(\frac{641}{5}\right)}{10}} = \sqrt{\frac{641}{25}} = \frac{\sqrt{641}}{5} m/s$

We can also calculate the velocity after 2 second by using the equation of motion as follows-

$\vec{v}_0 = 2\hat{i} + 3\hat{j} + 4\hat{k} \, m/s, \, \vec{a} = \frac{\hat{i}-\hat{j}-2\hat{k}}{10} m/s^2$

$t = 2$ sec., $\vec{v} = ?$

$\vec{v} = \vec{v}_0 + \vec{a}t$

$\vec{v} = (2\hat{i} + 3\hat{j} + 4\hat{k}) + \left(\frac{\hat{i}-\hat{j}-2\hat{k}}{10}\right)2$

or $\quad \vec{v} = (2\hat{i} + 3\hat{j} + 4\hat{k}) + \left(\frac{\hat{i}-\hat{j}-2\hat{k}}{5}\right)$

$= \frac{1}{5}(11\hat{i} + 14\hat{j} + 18\hat{k})$

$\therefore \quad |\vec{v}| = \frac{1}{5}\sqrt{121 + 196 + 324} = \frac{1}{5}\sqrt{641}$

EXAMPLE 22. A 2 kg ball when falls at rest through a height of 20 m acquires a speed of 10 m/s. Find the work done by air resistance (take $g = 10 \, m/s^2$).

APPROACH Since, the initial and final velocities of the ball are provided, therefore, we can easily apply the work energy theorem in this problem.

The forces acting on the ball are:

1. Gravitational force $m\vec{g}$, in vertically downward direction

2. Air resistance $\vec{R}$, in vertically upward direction.

The magnitude of displacement of the ball in downward direction is

$s = 20 \, m$

If, W_g and W_R are the work done by gravitational force and air resistance respectively, then net work done on the ball-

$W_{net} = W_g + W_R$

Since, the ball falls at rest and, finally, acquires the velocity of 10 m/s after the given displacement, therefore Initial kinetic energy, $K_i = \frac{1}{2}mv_i^2 = 0$

Final kinetic energy of the ball,

$K_f = \frac{1}{2}mv_f^2$, with $m = 2kg$,

$v_f = 10 \, m/s$.

Now, to find the work done by air resistance (W_R), substitute these values in work energy theorem-

$W_{net} = K_f - K_i \qquad \ldots (1)$

SOLUTION Work done by gravitational force,

$W_g = m\vec{g}.\vec{s} = mgs$

$= (2kg)(10 \, m/s^2)(20m) = 400 \, J$

Work done by air resistance =?

Initial kinetic energy, $K_i = 0$

Final kinetic energy, $K_f = \frac{1}{2}mv_f^2 = \frac{1}{2}(2kg)(10 \, m/s)^2$

$= 100 \, J$

Substituting these values in equation (1), we get

$W_{net} = K_f - K_i$

$W_g + W_R = 100J - 0$

or $\qquad 400 \, J + W_R = 100 \, J$

or $\qquad W_R = -300 \, J$

Here, negative sign indicates that the energy is transferred from ball to air.

EXAMPLE 23. A block of mass $m = 10 \, kg$ is projected up an inclined plane from its foot with a speed of 20 m/s as shown in the figure. The coefficient of kinetic friction μ_k between the block and the plane is 0.5. Find the distance traveled by the block on the plane before it stops first time.

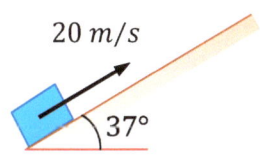

APPROACH As initial and final velocities of the block are provided, therefore, we use work energy theorem to find the distance travelled by the block.

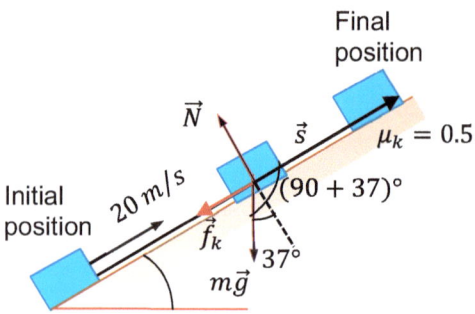

The forces acting on the ball are:
1. Gravitational force $m\vec{g}$, in vertically downward direction
$$mg = 10kg \times 10 \, m/s^2 = 100 \text{ newton}$$
2. Normal reaction $\vec{N}$, perpendicular to the inclined plane.
$$N = mg \cos 37° = (100 \text{ newton}) \times \left(\frac{4}{5}\right) = 80 \text{ newton}$$
3. Force of kinetic friction f_k, opposite to displacement vector $\vec{s}$
$$f_k = \mu_k N = (0.5)(80 \text{ newton}) = 40 \text{ newton}$$
Since, the normal reaction $\vec{N}$, is always perpendicular to $\vec{s}$, therefore the work done by it in above displacement is zero.
i.e., $W_N = 0$
The work done by gravitation force i.e., by weight, in displacement $\vec{s}$
$$W_g = m\vec{g}.\vec{s} = mgs \cos(90 + 37)°$$
$$= -mgs \sin 37° = -(100 \text{ newton}) s \left(\frac{3}{5}\right)$$
$$= -(60 \text{ newton}) s$$
The work done by force of kinetic friction $\vec{f_k}$, in displacement $\vec{s}$
$$W_{f_k} = \vec{f_k}.\vec{s} = (40 \text{ newton}) s \cos(180°)°$$
$$= -(40 \text{ newton}) s$$
The block starts from initial position with speed $v_1 = 20 \, m/s$ and stops at final position.
∴ Initial kinetic energy of the block: $K_i = \frac{1}{2} m v_1^2 = 2000 J$
Final kinetic energy of the block: $K_f = 0 J$
Now, to find displacement s, substitute above values in work energy theorem, $W_{net} = K_f - K_i$, with $W_{net} = W_g + W_N + W_{f_k} = -(60 \text{ newton}) s + 0 - (40 \text{ newton}) s$
$$= -(100 \text{ newton}) s$$
SOLUTION $W_{net} = K_f - K_i$
$$\Rightarrow -(100 \text{ newton}) s = 0 J - 2000 J$$
or $\quad s = \frac{2000 \, J}{100 \text{ newton}} = 20 \, m$

EXAMPLE 24. A block of mass m is attached to one end of a coiled spring of force constant k. The other end of the spring is fixed and the block can slide on a rough horizontal surface, where the coefficient of friction is μ_k. The block is held against the spring force compressing the spring by a distance x_0. The spring force in this position is more than force of limiting friction. Find the speed of the block when it passes the equilibrium position, when released.

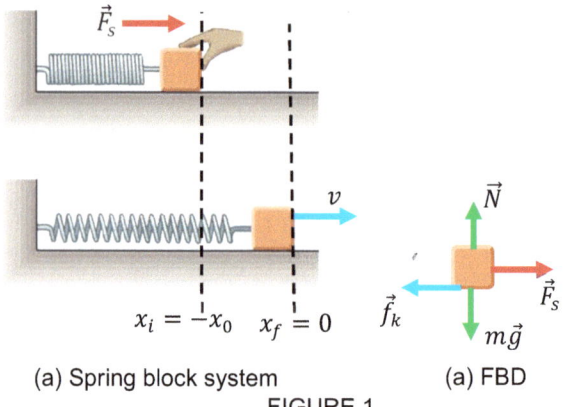

(a) Spring block system (a) FBD
FIGURE 1

APPROACH Apply work energy theorem to calculate the velocity of the block at equilibrium position.
$$W_{total} = K_f - K_i \quad \ldots (1)$$
SOLUTION Before reaching the equilibrium position, the forces acting on the block are-
(i) Gravitational force $m\vec{g}$ in vertically downward direction
(ii) Normal reaction $\vec{N}$, in vertically upward direction
(iii) Spring force $\vec{F_s}$, in positive x direction
(iv) Kinetic friction f_k, in negative x direction.
These forces are shown in FBD of the block (Fig. 1b)
Net work done,
$$W_{total} = W_{grav} + W_N + W_{spring} + W_{friction}$$
Since, $m\vec{g}$ and $\vec{N}$ are perpendicular to displacement vector, therefore the work done by these forces is zero, i.e.,
$$W_{grav} = W_N = 0$$
Work done by spring force, $W_{spring} = \frac{1}{2} k x_0^2$
and the work done by kinetic friction, $W_k = -\mu_k mg x_0$
∴ $\quad W_{total} = 0 + 0 + \frac{1}{2} k x_0^2 - \mu_k mg x_0$
or $\quad W_{total} = \frac{1}{2} k x_0^2 - \mu_k mg x_0$
Initial kinetic energy of the spring block system when the spring is compressed by distance x_0, $K_i = 0$
If block passes the equilibrium position with velocity v_0, then we have, $K_f = \frac{1}{2} m v_0^2$
Substituting these values in equation (1), we get
$$W = K_2 - K_1$$
$$\Rightarrow W_F + W_f = K_2 - K_1$$
$$\Rightarrow \frac{1}{2} k x_0^2 - \mu mg x_0 = \frac{1}{2} m v_0^2 - 0$$
$$\Rightarrow v_0 = \sqrt{(k x_0^2 - 2\mu mg x_o)}$$

EXAMPLE 25. A ball of mass m is suspended from a spring of force constant k. it is held to keep the spring in its relaxed length as shown in the Fig.1.

FIGURE 1

(a) The applied force is decreased gradually so that the ball moves downward at negligible speed. How far below the initial position will the ball stop?

(b) The applied force is removed suddenly. How far below the initial position, will the ball come to an instantaneous rest?

(a) **APPROACH** If the ball always remains in equilibrium, then during the whole motion, upward applied force and upward spring force ($F_{spring} = kx$) together balances the weight of the ball, i.e.,
$$F_{spring} + F_{applied} = mg \quad \ldots (1)$$
As ball moves downward, the expansion in spring (x) increases gradually and as a result the spring force (kx) also increases gradually. To keep the ball in equilibrium, we have to decrease the applied force gradually. At a certain expansion of spring, the spring force just balances the weight of the ball. At this stage, $F_{spring} = kx = mg$.
Therefore, from equation (1), we have
$$mg + F_{applied} = mg$$
or $\qquad F_{applied} = 0$

At this stage the ball comes at rest.
The initial and final positions and free body diagram of the ball at any intermediate position are shown Fig. 2.

SOLUTION If x_0 is the expansion of spring in equilibrium position when $F_{applied} = 0$, then,
$$kx_0 = mg$$
or $\qquad x_0 = \dfrac{mg}{k}$

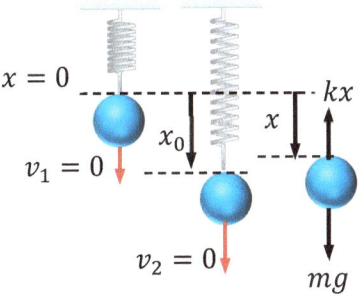
FIGURE 2

(b) **APPROACH** In the previous situation the applied force was decreased gradually keeping the ball everywhere in equilibrium. Now, if the applied force is removed suddenly, the ball will accelerate downwards. As the ball moves, the increase in spring extension increases the upward force, due to which acceleration decreases until extension becomes x_0. At this extension, the ball will acquire its maximum speed and it will move further downward due to the inertia of motion. When extension becomes more than x_0 spring force becomes more than the weight (mg) and the ball decelerates and ultimately stops at a distance x_m below the initial position. The initial position, the final position, and the free body diagram of the ball at some intermediate position when spring extension is x are shown in the Fig.3. The forces acting on the ball, are-
(i) gravitational force $F_{grav} = mg \downarrow$ (downwards)
(ii) spring force $F_{spring} = kx \uparrow$ (upwards)

Now, by applying work energy theorem, we can find the maximum extension (x_m) in the spring.

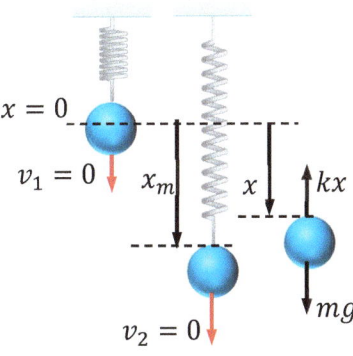
FIGURE 3

SOLUTION
Initial kinetic energy of the ball: $K_i = 0$
Final kinetic energy of the ball: $K_f = 0$
Work done by gravity, $W_{grav} = mgx_m$
Work done spring force, $W_{spring} = \dfrac{1}{2}kx_m^2$
Therefore, from work energy theorem, we get-
$$W_{total} = K_f - K_i$$
or $\qquad W_{grav} + W_{spring} = 0 - 0$
or $\qquad mgx_m + \dfrac{1}{2}kx_m^2 = 0$
or $\qquad x_m = 2mg/k$

EXAMPLE 26. In following figure, find how much m will rise if $4m$ falls away. Blocks are at rest and in equilibrium.

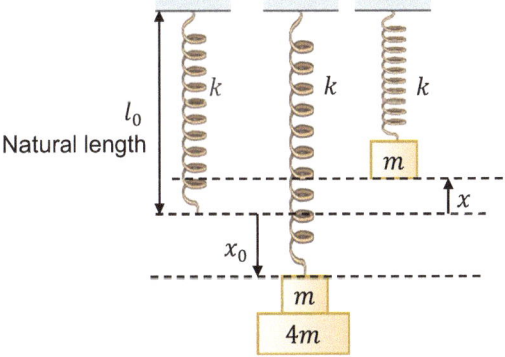

APPROACH Since all forces, initial and final kinetic energies can be easily calculated, therefore apply work energy theorem to find the distance above which block of mass m rise.

SOLUTION Let l_0 is the natural length of the spring and x_0 is the expansion in spring in equilibrium, then
$$kx_0 = 5mg$$
or $\qquad x_0 = \dfrac{5mg}{k}$

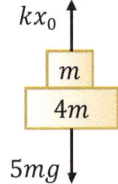

Now, suppose on falling of $4m$ block, the displacement of m is x above the natural length of spring, then by work energy theorem, we have
$$w_g + w_{sp} = k_f - k_i$$
If the displacement of block above the natural length of spring is x, then

$\underbrace{-mg\left(\dfrac{5mg}{k} + x\right)}_{\text{work done by gravity}} + \underbrace{\dfrac{1}{2}k\left(\dfrac{25\,m^2g^2}{k^2}\right)}_{\substack{\text{work done by spring in}\\\text{regaining it's original length}}}$
$+ \underbrace{\left(-\dfrac{1}{2}kx^2\right)}_{\substack{\text{work done by spring in}\\\text{extra compression } x}} = 0$

or $\quad \dfrac{1}{2}kx^2 + mgx - \dfrac{15m^2g^2}{2k} = 0$

or $\quad x = \dfrac{3mg}{k}$

Therefore, the displacement from initial position is
$\dfrac{5mg}{k} + \dfrac{3m}{k} = \dfrac{8mg}{k}$

EXAMPLE 27. A block of mass $m = 0.5\,kg$ slides from the point A on a horizontal track with an initial speed of $v_i = 3\,m/s$ towards a weightless horizontal spring of length $1\,m$ and force constant $k = 2\,N/m$. The part AB of the track is frictionless and the part BD has the coefficients of static and kinetic friction as 0.22 and 0.2 respectively. If the distance AB and BC are $2\,m$ and $2.14\,m$ respectively, find the total distance through which the block moves before it comes to rest completely (Take $g = 10\,m/s^2$).

$v_i = 3\,m/s$

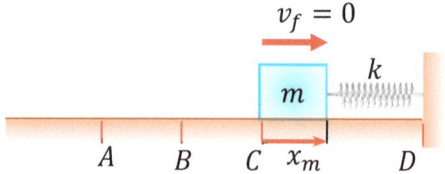

APPROACH Since all forces and initial and final kinetic energies can be easily calculated, therefore apply work energy theorem to find the total distance travelled by the block.

SOLUTION Since, $m\vec{g}$ and $\vec{N}$ are always perpendicular to displacement vector, therefore the work done by these forces will be zero, i.e.,
$$W_{grav} = W_N = 0$$
The portion AB of the track is smooth, therefore, the block reaches B with same velocity v_i. Afterward force of kinetic friction starts opposing its motion. As the block passes the point C the spring force also starts opposing its motion in addition to the force of kinetic friction. The work done by these forces decrease the kinetic energy of the block and stop the block momentarily at a distance x_m after the point C (see following figure).

Initial kinetic energy of the block at position-A is
$$K_i = \dfrac{1}{2}mv_i^2 = \dfrac{1}{2}\times 0.5 \times 9 = 2.25J$$
Final kinetic energy of the block, when, the spring is fully compressed
$$K_f = \dfrac{1}{2}mv_f^2 = 0J$$
Work done by the frictional force before the block stops is
$$W_{fr} = -\mu_k mg(BC + x_m)$$
$$= -0.2 \times 0.5 \times 10(2.14 + x_m)$$
$$= -(2.14 + x_m)J$$
Work done by the spring force before the block stops is
$$W_{spring} = \int_{x=0}^{x_m} kx\,dx = \dfrac{1}{2}kx_m^2 = \dfrac{1}{2}(2)(x_m^2) = x_m^2$$
Now, from work energy theorem, we have
$$W_{total} = K_f - K_i$$
or $\quad W_{fr} + W_{spring} = 0 - 2.25$
$\Rightarrow -(2.14 + x_m) + x_m^2 = -2.25$
On solving for x_m, we get
$\Rightarrow x_m = 0.1m$

The motion of block after it stops momentarily at maximum compression x_m depends upon the condition whether the spring force is more than or less than the force of limiting friction. If the spring force in this position is more than the force of limiting friction, then, the block will move back and if the spring force in in this position is less than the force limiting friction, then the spring force will not be able to move the block in backward direction and it stops permanently at the compressed position of spring.

Spring force F_s at final position of the block-
$$F_s = kx_m = 0.2N$$
The force of limiting friction
$$f_{s,max} = \mu_s mg = 1.1N$$
Here, it is clear that-
$$F_s < f_{s,max}$$
Therefore, the spring force won't be able to move back the block, and the block remains stationary permanently at the maximum compression state of spring.
The total distance travelled by the block
$= AB + BC + x_m = 2m + 2.14m + 0.1 = 4.24m$

EXAMPLE 28. (a) A 2 kg block situated on a smooth fixed incline is connected to a spring of negligible mass, with spring constant $k = 100\,Nm^{-1}$, via a frictionless pulley. The block is released from rest when the spring is unstretched. How far does the block move down the incline before coming (momentarily) to rest? What is its acceleration at its lowest point?

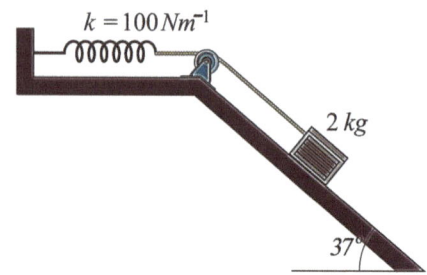

(b) The experiment is repeated on a rough incline. If the block is observed to move 0.20 m down along the incline before it comes to instantaneous rest, calculate the coefficient of kinetic friction.

APPROACH As block is released from rest and finally also comes at rest, therefore we can easily calculate the initial and final kinetic energies. The forces acting on the block can also be calculated easily. Therefore, to calculate the distance covered by the block, apply work energy theorem.

SOLUTION (a) *Applying work-energy theorem*

$$mgs \sin 37° = \frac{1}{2}ks^2$$

$$2 \times 10 \times s \times \frac{3}{5} = \frac{1}{2} \times 100 \times s^2 \Rightarrow s = 0.24\ m$$

Acceleration at its lowest point

$$a = \frac{ks - mg \sin 37°}{m} = \frac{100 \times 0.24 - 2 \times 10 \times \frac{3}{5}}{2} = 6\ m/s^2$$

or $\quad a = 6\ m/s^2$

(b) By work energy theorem, we have

$$W_c + W_{nc} = K_2 - K_1$$

⇒ work done by spring + work done by gravity + work done by friction = $K_2 - K_1$

or $\left(-\frac{1}{2}ks^2\right) + mgs \sin 37° - \mu mg \cos 37° \times s = 0 - 0$

or $\quad \left(-\frac{1}{2}ks\right) + mg \sin 37° = \mu mg \cos 37°$

or $\quad -\frac{1}{2} \times 100 \times 0.20 + 2 \times 10 \times \frac{3}{5} = \mu \times 2 \times 10 \times \frac{4}{5}$

or $\quad \mu = \frac{1}{8}$

13.2. WORK AND KINETIC ENERGY IN COMPOSITE SYSTEMS

The work energy theorem is applicable only to bodies that we can represent as *particles*—that is, as moving point masses. Here, we consider some examples of complex systems that have to be represented as many particles with different motions.

1. Suppose, a boy stands on frictionless roller skates on a level surface, facing a rigid wall (Fig. 1). He pushes against the wall, which makes him move to the right. The forces acting on him are his weight $\vec{W}$, the upward normal forces $\vec{N_1}$ and $\vec{N_2}$ exerted by the ground on his skates, and the horizontal force $\vec{F}$ exerted on him by the wall. There is no vertical displacement, so $\vec{W}$, $\vec{N_1}$ and $\vec{N_2}$ do no work. Force $\vec{F}$ accelerates him to the right, but the point of application of force, i.e., the parts of his body where that force is applied (the man's hands) do not move while the force acts. Thus, the force also does no work. Where, then, does the boy's kinetic energy come from?

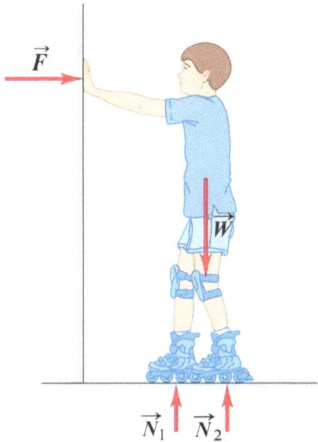

FIGURE 1. The external forces acting on a skater pushing off a wall. The work done by these forces is zero, however the skater's kinetic energy changes.

The explanation is that it's not adequate to apply the particle model for the boy because as he extends his arms, is a *deformable object* and the particle model is not applicable for a deformable object. The boy, has an *internal source of energy*. Because he is a living object, he has an internal store of chemical energy that is available through metabolic processes. *Always remember that a system can gain kinetic energy without any work being done if it can transform some other energy into kinetic energy*. In this case, the boy transforms chemical energy into kinetic energy with the help of external force on him. The same is true if you jump straight up from the ground. The ground applies an upward force to your feet, but that force does no work because the point of application—the soles of your feet—has no displacement while you're jumping. Instead, your increased kinetic energy comes via a decrease in your body's chemical energy. A brick cannot jump or push off from a wall because it cannot deform and has no usable source of internal energy.

2. Figure 2 shows another example. An engine increases the speed of a car with four-wheel drive (all four wheels are made to turn by the engine). During the acceleration, the engine rotates the tyres of the car, by converting the internal chemical energy stored in the fuel to mechanical energy and causes the tyres to push backward on the road surface. This push produces frictional forces $\vec{f}$ that act on each tyre in the forward direction. The net external force $\vec{F}$ from the road, which is the sum of these frictional forces, is responsible for acceleration of the car. As a result of which the kinetic energy of the car increases. However, net static friction $\vec{F}$ does no work on the car in ground frame because there is no relative slipping between the tyres and ground i.e., the point of application of static frictional forces on each tyre always remains at rest with respect to the ground. Thus, although

the work done by external force on car is zero, it is responsible for gaining in kinetic energy.

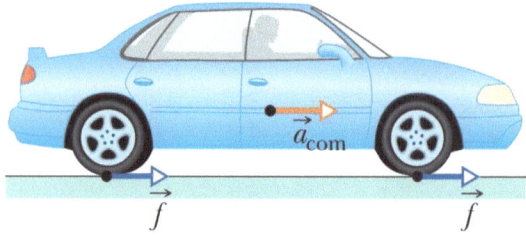

FIGURE. 2

☞ In all the above cases, we cannot apply **work energy theorem**. Work energy theorem is only applicable for the situations where the system can be considered as a point mass.

KEYPOINT 1. Work energy theorem is only applicable for the situations where the system can be considered as a point mass.
2. An external force can change the kinetic energy or potential energy of an object without doing work on the object—that is, without transferring energy to the object. Instead, the force is responsible for transfers of energy from one type to another inside the object.

IMPORTANT POINTS

- Work is a scalar (it is a dot product of two vectors) having dimension $[ML^2T^{-2}]$ and SI unit $N.m$ [or joule (J)]. A practical unit of work in atomic and nuclear physics is electron volt (eV)
$$1eV = 1.6 \times 10^{-19} J$$
- In case of a charged particle in a magnetic field, the direction of the magnetic force $\vec{F}\,[= q(\vec{v} \times \vec{B})]$ is always perpendicular to velocity, therefore the work done by this force is always zero.
- In a circular motion, work done by centripetal (i.e., radial or central) force is always zero (as the radial force is always perpendicular to the tangential displacement)); so, from work energy theorem ($W_{net} = K_f - K_i$), the kinetic energy and speed remains constant (while velocity and momentum changes due to change in direction) and motion is uniform. However, if in addition to radial force there is also a tangential force, work done by tangential force will not be zero and so, KE will not remain constant (∵ $W = \Delta KE$, will be discussed later) and motion will not be uniform.
- *Work depends on frame of reference.* With change of frame of reference (inertial) force does not change while displacement may change; so, the work done by a force will be different in different frames, e.g.,
a) If a porter with a suitcase on, his head moves up a staircase, work done by the upward lifting force on suitcase, relative to him, will be zero (as displacement relative to him is zero). while relative to a person on the ground will be $\vec{F}.\vec{s} = mgh$ (as $F = mg$ and $s = h$)
b) If a person is pushing a box inside a moving train, the work done in the frame of train will be $(\vec{F}.\vec{s})$ while in the frame of earth will be $\vec{F}.(\vec{s} + \vec{s_0})$, where $\vec{s_0}$ is the displacement of the train relative to the ground.

14. CHECKPOINT 3

1. •A $6.0\ kg$ block initially at rest is pulled to the right along a horizontal frictionless surface by a constant force of $12\ N$, as shown in the following figure. Find the speed of the block after it has moved $3.0\ m$

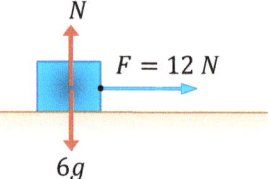

2. •A ball of mass m is thrown in air with speed v_1 from a height h_1 and it is caught at a height $h_2 > h_1$ when its speed becomes v_2. Find the work done on the ball by the air resistance.

3. ••A body of mass $1\ kg$ thrown upwards with a velocity of $10\ m/s$ comes to rest (momentarily) after moving up $4m$. Find the work done by air drag in this process is (Take $g = 10\ m/s^2$).

4. ••A particle of mass m is projected with velocity v_0 at an angle θ_0 with horizontal. During the period, when the particle descends from highest point to the position where its velocity vector makes an angle $\theta_0/2$ with horizontal. Find the work done by the gravity.

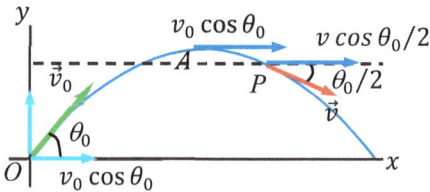

5. ••A box of mass $123\ kg$, initially at rest on a horizontal surface, is pushed in a straight line with a constant force $(\vec{F})$ of magnitude $560\ N$ over a distance of $7.0\ m$. Over this distance, the speed of the box changes from zero to $4.0\ m/s$. The frictional force between the box and the surface cannot be ignored. Determine the coefficient of kinetic friction between the box and the surface.

6. ••Following figure shows a cord attached to a cart that can slide along a frictionless horizontal rail

aligned along an x axis. The left end of the cord is pulled by a constant force of $25N$ over an ideal pulley at cord height $h = 1.20\,m$. If during the motion, the cart slides from $x_1 = 3.00\,m$ to $x_2 = 1.00\,m$, then find the change in the kinetic energy of the cart.

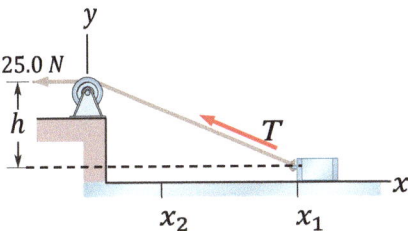

7. • Consider the following true-false questions:

(a) *True or False?*
A boy jumps up into the air by applying a force downward on the ground. The work-kinetic energy theorem $W = \Delta K$ can be applied to the boy to find the speed with which he leaves the ground.

(b) *True or False?*
A balloon is compressed uniformly from all sides. Because there is no displacement of the balloon's centre of mass, no work is done on the balloon.

15. WORK OF CONSTRAINT FORCES

Constraint Forces are the forces that the constraining object exerts on the object to make it follow the motional constraints (For example, tension in connected pulleys). As these forces are always internal to the system, therefore, their net work on entire system is always zero. i.e., $\sum W_{constraint} = 0$.

The method of constraint work is very useful for problems involving several connected rigid bodies. The step wise procedure is given in following example-

EXAMPLE 29. Find the constrained relations between displacements, velocities and accelerations of two blocks of masses m_1 and m_2 (Fig. 1a) by using the method of work done by constraint forces.

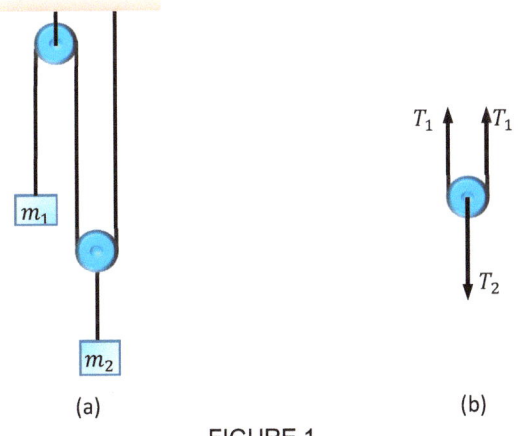

FIGURE 1.

Step–1. To apply this method, write constraint forces on objects under consideration. Here tensions in the strings are constraint forces. Suppose, the tension in string connected to mass m_2 is T_2 and in string connected to mass m_1 is T_1.

Step–2. Set the fixed point as origin and displacements away from it as positive. In the adjoining diagram, the displacements in downward direction are positive.

Step 3. Displace each of the movable bodies in $+ve$ direction (away from a fixed point) on the free-body diagram. by distances $s_1, s_2, s_2...$ etc. Here we need not bother whether these displacements are physically possible or not. Automatically the analysis will tell you the relationship between them. In the adjoining figure, the displacement of mass m_1 in positive direction is s_1 and that of m_2 is s_2.

Step–3. Find the work done by tension on each of the bodies. The sum of all these works should be zero.

In adjoining diagram, we have assumed that both bodies are moving in downward direction, therefore, the work done by T_1,

$W_1 = -T_1 s_1$ ($\because T_1$ and s_1 are oppositely directed)

and work done by T_2,

$W_2 = -T_2 s_2$ ($\because T_2$ and s_2 are oppositely directed)

$\therefore$ net work done by forces T_1 and T_2 is

$W = W_1 + W_2 = -T_1 s_1 - T_2 s_2$

But we know that, the net work done by constraint forces over entire system is always zero

$\therefore \quad W = 0$

or $\quad -T_1 s_1 - T_2 s_2 = 0$

or $\quad T_1 s_1 + T_2 s_2 = 0$ ... (1)

Now, considering the FBD of the massless movable pulley (Fig. 1b). From, Newton's second law, we have

$\sum F = ma$

$T_2 - 2T_1 = 0$ ($\because$ mass of the pulley is zero)

or $\quad T_2 = 2T_1$ ... (2)

Substituting this value of T_2, in Eq. (1), we get

$T_1 s_1 + 2T_1 s_2 = 0$

or $\quad s_1 + 2s_2 = 0$

Similarly, we can show that the velocities of m_1 and m_2 will be related as

$v_1 + 2v_2 = 0$

and their accelerations are related as

$a_1 + 2a_2 = 0$

EXAMPLE 30. *In the figure shown, the ring starts moving down from rest. What will be the relation between the velocity of the ring and the velocity of the block at any position? What will be the distance that the ring moves before coming to rest?*

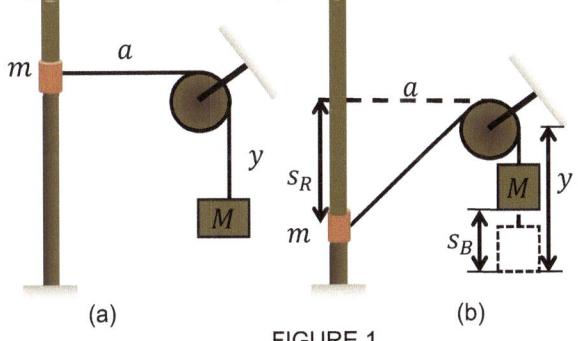

FIGURE 1

APPROACH To find the relationship between the velocities of the block and the ring, we apply the concept of work done by constraint force. Here, the constraint force is the tension in the string, therefore, the total work done by tension in the given system must be zero.

SOLUTION Suppose, ds_B and ds_R are very small downward displacements of block and ring respectively, at angle θ, as shown in Fig. 2 (displacement is assumed to be very small, so that the angle made by the string does not change appreciably), then the small work done by tension on block,
$$dW_B = -T.ds_B$$
Work done by tension on the ring,
$$dW_R = -Tds_R\cos\theta$$
Total work done by constraint force on the entire system,
$$dW = dW_B + dW_R$$
Substitute the values of dW_B and dW_R in above equation, we get
$$dW = -T.ds_B - Tds_R\cos\theta$$
According to principle of work by constraint forces,
$$dW = 0$$
i.e., $-T.ds_B - Tds_R\cos\theta = 0$
or $\quad ds_B + ds_R\cos\theta = 0$
Dividing both sides by small time dt, we get
$$\frac{ds_B}{dt} + \frac{ds_R}{dt}\cos\theta = 0$$
or $\quad v_B + v_R\cos\theta = 0$ ... (1)

☞ Here it should be noted, that, the relationship between the small displacements is similar to that of velocities because we have divided the entire relationship by small time interval dt. But to obtain the relationship between accelerations, we have to differentiate this expression which will also involve the derivative of $\cos\theta$ because θ is also a variable.

Suppose, ring moves by distance s_R and block by s_B before coming to rest, then work done by gravity,
$$W_g = mgs_R - Mgs_B \quad ... (2)$$
As the work done by constraint force i.e., tension is zero, therefore the only force which does work is the gravitational force.
Applying, work energy theorem, we get
$$W_g = \Delta K_R + \Delta K_B \quad ... (3)$$
Here, ΔK_R is the change in kinetic energy of ring and ΔK_B is the change in kinetic energy of block.
Using Eq. (2) in Eq. (3), we get
$$mgs_R - Mgs_B = \frac{1}{2}mv_R^2 + \frac{1}{2}Mv_B^2$$

☞ The interesting thing to note here is that even if mass of the ring is less than that of the block, it is moving downward.

At the position of rest, $v_B = 0$. So, from the equation of constrained motion, v_R is also 0.
$$mgs_R - Mgs_B = 0$$
or $\quad ms_R - Ms_B = 0$
or $\quad \frac{s_R}{s_B} = \frac{M}{m}$ ... (4)
Also, from the geometry, the original length of the string is-
$$a + y = l$$
or $\quad \sqrt{a^2 + s_R^2} + y - s_B = a + y$
or $\quad \sqrt{a^2 + s_R^2} - s_B = a$
or $\quad \sqrt{a^2 + s_R^2} = a + s_B$
Squaring both sides, give
$$a^2 + s_R^2 = a^2 + s_B^2 + 2as_B$$
or $\quad s_R^2 = s_B^2 + 2as_B$
Dividing both sides by s_B^2, we get
$$\left(\frac{s_R}{s_B}\right)^2 = 1 + \frac{2a}{s_B} \quad ... (5)$$
Using Eq. (4), in Eq. (5), we get
$$\left(\frac{M}{m}\right)^2 = 1 + \frac{2a}{s_B}$$
or $\quad \frac{2a}{s_B} = \frac{M^2}{m^2} - 1$
or $\quad \frac{2a}{s_B} = \frac{M^2 - m^2}{m^2}$
or $\quad s_B = \frac{2am^2}{M^2 - m^2} \quad ... (6)$
Now, substitute this value in (4), you get
$$s_R = \frac{M}{m}s_B = \frac{M}{m}\left(\frac{2am^2}{M^2 - m^2}\right)$$
or $\quad s_R = \frac{2amM}{M^2 - m^2}$

EXAMPLE 31. In following figure, $m_A = 1 kg$, $m_B = 2\,kg$, $m_C = 10\,kg$. If there is a relative slipping between A and B, then find the velocities of A, B & C when C has descended by $2m$ Calculate the time taken by C to cover this distance.

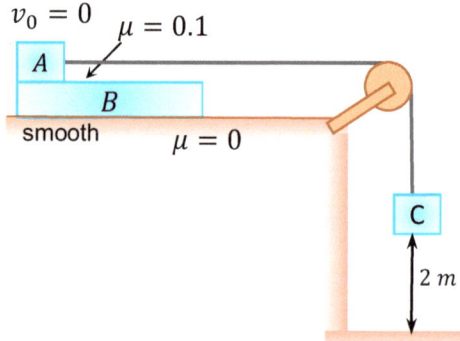

APPROACH Apply work energy theorem for each individual body shown in adjoining figure.
SOLUTION *For block A*:
The forces acting on A, are-
(i) Gravitational force $1g$ newton in vertically downward direction
(ii) Normal reaction $N_1 = 1g = 10$ newton, in vertically upward direction
(iii) Tension T, in right direction
(iv) Force of kinetic friction $f_k = \mu mg = (0.1)(1kg)(10\,m/s^2) = 1$ newton, in left direction.
Since, normal reaction and gravitational forces are perpendicular to displacement vector, therefore, their work done on block A will be zero.
Therefore, net work done on block A,
$$W_A = T \times 2 - f_k \times 2 = T \times 2 - 1 \times 2 \quad ... (1)$$

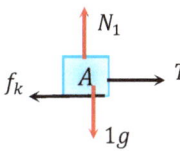

For block C:
The forces acting on C, are-
(i) Gravitational force $10g$ newton = 100 newton in vertically downward direction
(ii) Tension T, in vertically upward direction
Therefore, net work done on block C,
$$W_C = 100 \times 2 - T \times 2 \quad \ldots (2)$$

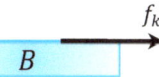

Adding (1) and (2), we get
$$W_A + W_B = 100 \times 2 - 1 \times 2 = 198\,J$$
For system $A + C$, applying work energy theorem, we get
$$W_A + W_B = (K_f - K_i)_A + (K_f - K_i)_B$$
or $\quad 198\,J = \left(\frac{1}{2} \times 1 \times v^2 - 0\right) + \left(\frac{1}{2} \times 10 \times v^2 - 0\right)$
or $\quad 198\,J = \frac{1}{2} \times 11\, v^2$
i.e., $\quad v^2 = \frac{396}{11} = 36$
or $\quad v = 6\, m/s$

Therefore, the velocities of each A and C will be $6\, m/s$.
Calculation of time taken by C, in travelling $2m$ distance:
$v_0 = 0, s = 2m, v = 6\,m/s, t = ?$
By $v^2 = v_0^2 + 2as$, we have
$a = \frac{v^2 - v_0^2}{2s} = \frac{36 - 0}{2 \times 2} = 9\,m/s^2$
Now, applying $v = v_0 + at$, we get
$t = \frac{v - v_0}{a} = \frac{6 - 0}{9} = \frac{2}{3}$ sec.

Now, for block B

Acceleration of the block in horizontal direction,
$a_B = \frac{f_k}{m_B} = \frac{1}{2} = 0.5\,m/s^2$
Initial velocity of B, $v_{0B} = 0, t = 2/3$ s
$\therefore s = v_0 t + \frac{1}{2} a t^2$ gives
$s = 0 + \frac{1}{2} \times \frac{1}{2} \times \frac{4}{3} = \frac{1}{9} m$
Therefore, from work energy theorem, we have
$$W_C = K_f - K_i = \frac{1}{2} m_C v^2 - 0$$
or $\quad \vec{F}.\vec{s} = 1 \times \frac{1}{9} = \frac{1}{2} \times 2 \times v^2 - 0 \quad \Rightarrow v_B = \frac{1}{3} ms^{-1}$

16. VIRTUAL WORK

The principle of virtual work was proposed by the Swiss mathematician Jean Bernoulli in the eighteenth century. It provides an alternative method for solving problems involving the equilibrium of a particle, a rigid body, or a system of connected rigid bodies. Before we discuss this principle, however, we must first define the virtual work produced by a force and by a couple of moment (will be discussed later in the chapter 'Dynamics of a Rigid Body').

16.1. VIRTUAL WORK OF A FORCE

The virtual work is the work done by an external force on a body due to a very small virtual (or imaginary) displacement of the body.
Here, the virtual displacement is a displacement that is assumed and does not actually exist.
The virtual work done by a force $\vec{F}$, having a virtual displacement $\delta\vec{r}$, is
$$dW = \vec{F}.d\vec{r} \quad \ldots (1)$$

16.2. VIRTUAL WORK OF A COUPLE

The virtual work of a couple of moment $\vec{\tau}$ (i.e., torque) acting on a rigid body is given by-
$$dW = \vec{\tau}.d\vec{\theta} \quad \ldots (2)$$
where $d\vec{\theta}$ is the small virtual angular displacement (expressed in radians) through which the body rotates.

16.3. PRINCIPLE OF VIRTUAL WORK

The principle of virtual work states that if a body is in equilibrium, then the algebraic sum of the virtual work done by all the external forces and couple moments acting on the rigid body, is zero for any virtual displacement of the body. Thus, $\sum dW = 0$.

FIGURE 1

For example, consider the free-body diagram of the particle (ball) that rests on the floor, Fig. 1. If we "imagine" the ball to be displaced downwards a virtual amount dy, then the weight does positive virtual work, $mg\,dy$ and the normal force does negative virtual work, $-N\,dy$. For equilibrium the total virtual work must be zero, so that-
$\sum dW = mg\,dy - N\,dy = 0$
$\Rightarrow \quad (mg - N)dy = 0$
$\because \quad dy \neq 0$
$\therefore \quad mg - N = 0$ (as required by applying $\sum F_y = 0$)

17. CONSERVATIVE AND NON-CONSERVATIVE FORCES

On the basis of work energy concept, we generally divide forces into two classes, conservative and nonconservative. The distinction between a conservative and a nonconservative force is based on the work the force does over a given path.

17.1. CONSERVATIVE (OR INTERNAL) FORCES

If work done by a force on a particle *moving between any two points,* depends only on the initial and final positions and hence is independent of the path taken between the two points, then the force is called a **conservative force**. Examples: Gravitational force, spring force, electric force, all central forces

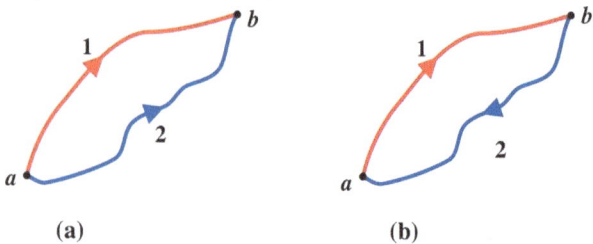

(a) (b)

Fig. 1 (a) A conservative force acts on a particle moving from point a to point b by following either path 1 or path 2. (b) A conservative force acts on a particle moving in a round trip from point a to point b along path 1 and then back to point a along path 2

With reference to the arbitrary paths of Fig.1a, we can write the first condition for a conservative force as:
$$W_{ab}(\text{path 1}) = W_{ab}(\text{path 2}) \quad \ldots (1)$$
i.e., the work done by a conservative force on a particle moving from *a* to *b* along path 1 is the same as from *a* to *b* along path 2.
Also, with reference to the arbitrary paths of Fig. 1b, we can write the second condition for a conservative force as:
$$\left.\begin{array}{l}W_{ab}(\text{path 1}) = -W_{ba}(\text{path 2}) \\ \text{or} \\ W_{ab}(\text{path 1}) + W_{ba}(\text{path 2}) = 0\end{array}\right\} \quad \ldots (2)$$
That is, the work done by a conservative force on a particle that moves in a round trip from *a* to *b* along path 1 and then from *b* to *a* along path 2 is zero. In other words:
The net work done by a conservative force on a particle that is moving around any closed path is zero.
Mathematically,
$$W = \oint \vec{F}(\vec{r}) \cdot d\vec{r} = 0 \quad \ldots (3)$$
(conservative force)
The circle symbol around the integral sign indicates that the integral is to be taken over a closed path.
For motion in one dimension, the above equation takes the form,
$$W = \oint F(x) dx = 0$$
Thus, for a conservative force, in a round trip, the work energy theorem gives
$$W = K_f - K_i = 0 \quad \text{or} \quad K_f = K_i$$
i.e., the particle will return to its starting point with the same kinetic energy it had when it started its motion.
☞ A conservative force can be a function only of position, and cannot depend on other variables like time or velocity.

We can show that the gravitational force is a conservative force. The gravitational force on an object of mass m near the Earth's surface is $m\vec{g}$, where $\vec{g}$ is a constant. The work done by this gravitational force on an object that moves a vertical distance h in upward direction, is
$$W_g = -mgh \quad \text{(see Fig. 2a)} \quad \ldots (1)$$
(force and displacement are oppositely directed)
If the same object falls by a vertical distance h, then the work done by gravitational force
$$W_g' = +mgh \text{ (see Fig. 2a)} \quad \ldots (2)$$
Now, suppose that instead of moving vertically upward, an object follows some arbitrary path in the xy plane, as shown in Fig. 2b. The object starts at a vertical height y_1 and reaches a height y_2, where $y_2 - y_1 = h$. To calculate the work done by gravity, we use:
$$W_g = \int_1^2 m\vec{g} \cdot d\vec{l} = \int_1^2 mg \cos\theta \, dl \quad \ldots (3)$$

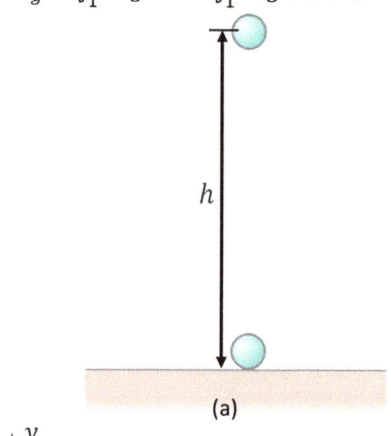

(a)

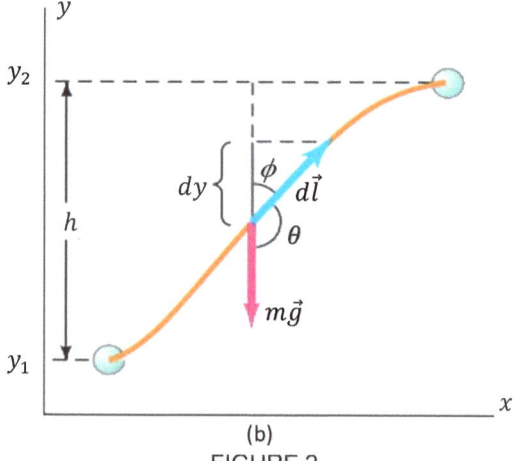

(b)
FIGURE 2

If the angle between $d\vec{l}$ and its vertical component dy is ϕ, then from Fig. 2b, we have-
$$\theta = 180° - \phi$$
$$\therefore \cos\theta = \cos(180° - \phi) = -\cos\phi$$
and $dy = dl \cos\phi$,
$$\therefore \quad W_g = -\int_1^2 mg \, dy = -mg(y_2 - y_1) \quad \ldots (4)$$
From Fig. 2b, $(y_2 - y_1) = h$, therefore from Eq. 4, we have
$$W_g = -mgh \quad \ldots (5)$$

Similarly, if object returns from y_2 to y_1, then work done by gravity
$$W_g' = +mgh \quad \ldots (6)$$
Now, if we compare Eq. (1) and (5) or Eq. (2) and (6). we find that the work done by gravity depends only on the vertical height and does not depend on the particular path taken! Hence, by definition, gravity is a conservative force.

In summary, the work done by a conservative force has four properties:
1. It can be expressed as the difference between the initial and final values of a *potential-energy* function.
2. It is reversible.
3. It is independent of the path of the body and depends on only the starting and ending points.
4. When the starting and ending points are the same, the total work is zero. When the *only* forces that do work are conservative forces, the total mechanical energy $E = K + U$ is constant.

17.2. NON-CONSERVATIVE (OR EXTERNAL) FORCES

Not all forces are conservative. For example, let us allow a book to slide across a table that is not frictionless, see Fig.3a. During the sliding, the kinetic frictional force does negative work on the book, slowing it by transferring energy from its kinetic energy to thermal energy of the book-table system. This energy transfer cannot be reversed. So, this force is not conservative. Therefore, all types of frictional forces are **non-conservative** forces. That is:

The work done by a non-conservative force on a particle that is moving between any two points depends on the path taken by the particle.

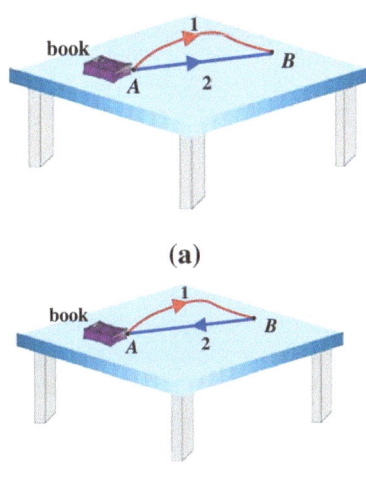

FIGURE 3. *(a) The work done by the force of friction depends on the path taken as the book is moved from A to B. (b) The work done by the force of friction in a round trip from point A to point B along path 1 and then back to point A along path 2 is not zero*

With reference to the arbitrary paths of Fig. 3a, we can write the first condition for a non-conservative force as:
$$W_{AB}(\text{path 1}) \neq W_{AB}(\text{path 2}) \quad \ldots (3)$$
(Non-conservative forces)

i.e., the work done by a non-conservative force on a particle moving from A to B along path 1 is always not the same along path 2.

Also, with reference to the arbitrary paths of Fig. 3b, we can write the second condition for a non-conservative force as:
$$\left. \begin{array}{c} W_{ab}(\text{path 1}) \neq -W_{ba}(\text{path 2}) \\ \text{or} \\ W_{ab}(\text{path 1}) + W_{ba}(\text{path 2}) \neq 0 \end{array} \right\} \quad \ldots (4)$$

That is, the work done by a non-conservative force on a particle that moves in a round trip from A to B along path 1 and then from B to A along path 2 is not zero.

☞ The work done by a nonconservative force *cannot* be represented by a potential-energy function. Some nonconservative forces, like kinetic friction or fluid resistance, cause mechanical energy to be lost or dissipated; a force of this kind is called a **dissipative force.** There are also nonconservative forces that *increase* mechanical energy. The fragments of an exploding firecracker fly off with very large kinetic energy, it is due to a chemical reaction of gunpowder with oxygen. The forces unleashed by this reaction are nonconservative because the process is not reversible. (The fragments never spontaneously reassemble themselves into a complete firecracker!)

17.3. IDENTIFICATION OF A PLANAR CONSERVATIVE FORCE

Suppose a given force is $\vec{F} = F_x \hat{i} + F_y \hat{j}$. To check whether the given planar force is conservative or not, we find partial derivatives $\frac{\partial F_x}{\partial y}$ and $\frac{\partial F_y}{\partial x}$.

Now, if $\frac{\partial F_x}{\partial y} = \frac{\partial F_y}{\partial x}$, then, the given force is a conservative force.

and if, $\frac{\partial F_x}{\partial y} \neq \frac{\partial F_y}{\partial x}$, then the given force is a non-conservative.

$$\vec{F} = \begin{cases} \text{conservative, if } \partial F_x/\partial y = \partial F_y/\partial x \\ \text{nonconservative, if } \partial F_x/\partial y = \partial F_y/\partial x \end{cases} \quad \ldots (1)$$

For example, consider the force, $\vec{F} = 2x^2y^2\hat{i} + xy^2\hat{j}$
On comparing above expression with, $\vec{F} = F_x\hat{i} + F_y\hat{j}$, we get-
$$F_x = 2x^2y^2, F_y = xy^2$$
Now, $\frac{\partial F_x}{\partial y} = 4x^2y, \frac{\partial F_y}{\partial x} = y^2$

Clearly, $\frac{\partial F_x}{\partial y} \neq \frac{\partial F_y}{\partial x}$,

Therefore, the given force is a non-conservative force.

KEYPOINT In case of non-conservative forces, work cannot be calculated without describing the path.

EXAMPLE 32. A force $\vec{F} = xy\hat{i} + xy\hat{j}$, displaces a particle from $(0,0)$ to $(2,2)$. Calculate the work required to take particle from $(0,0)$ to $(2,2)$ along path I and II.

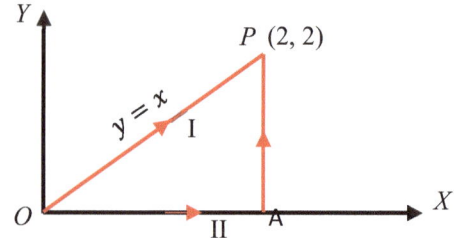

APPROACH Compare the given force with $\vec{F} = F_x\hat{i} + F_y\hat{j}$ and test for conditions-
$$\vec{F} = \begin{cases} \text{conservative, if } \partial F_x/\partial y = \partial F_y/\partial x \\ \text{nonconservative, if } \partial F_x/\partial y \neq \partial F_y/\partial x \end{cases} \quad \ldots (1)$$

SOLUTION Comparing, $\vec{F} = xy\hat{i} + xy\hat{j}$, with $\vec{F} = F_x\hat{i} + F_y\hat{j}$, we get
$F_x = xy$ and $F_y = xy$
$\therefore \frac{\partial F_x}{\partial y} = x \neq \frac{\partial F_y}{\partial x} = y$

So, the given force is non-conservative, hence the work done will be path dependent.
Now, work
$$dW = (F_x\hat{i} + F_y\hat{j}) \cdot (dx\hat{i} + dy\hat{j}) = F_x dx + F_y dy$$
or $dW = \vec{F} \cdot d\vec{r} = F_x dx + F_y dy$

Calculation of work along path I (along path I, $y = x$)
$\therefore \quad W = \int_0^2 F_x dx + \int_0^2 F_y dy = \int_0^2 xy dx + \int_0^2 xy dy$

With the help of $y = x$, converting first term of RHS as a function of x and second term as a function of y, we get-
$$W = \int_0^2 x(x) dx + \int_0^2 (y) y dy$$
or $W = \int_0^2 x^2 dx + \int_0^2 y^2 dy = \left[\frac{x^3}{3}\right]_0^2 + \left[\frac{y^3}{3}\right]_0^2 = \frac{16}{3} J$

Calculation of work along path II (along path II, y is 0 from $x = 0$ to 2 and x is 2 from $y = 0$ to 2.)
$W = \int_0^2 F_x dx + \int_0^2 F_y dy = \int_0^2 xy dx + \int_0^2 xy dy$
$= \int_0^2 x(0) dx + \int_0^2 (2) y dy$
$= 0 + 2\int_0^2 y dy = \left[\frac{y^2}{3}\right]_0^2 = \frac{4}{3} J$

EXAMPLE 33. Find an expression for the work done by a central force, which follow the inverse square law.

APPROACH According to the given problem, the central force* is following inverse square law i.e., $F \propto \frac{1}{r^2}$
or $F = \frac{k}{r^2}$, here, k is a proportionality constant.
In vector form, $\vec{F} = \frac{k}{r^2}\hat{r}$

Now, we have to show that, if an object moves under above force from point A to B, then work done by it is path independent.

SOLUTION Let us consider a central force $F = \frac{k}{r^2}$.
where, $\vec{r} = x\hat{i} + y\hat{j} + z\hat{k}$, therefore $d\vec{r} = dx\hat{i} + dy\hat{j} + dz\hat{k}$ and $r = \sqrt{x^2 + y^2 + z^2}$.
Now, $\vec{F} = \frac{k}{r^2} = \frac{k}{r^2}\hat{r} = \left(\frac{k}{r^2}\right)\left(\frac{\vec{r}}{r}\right) = \frac{k\vec{r}}{r^3}$
Substituting the valued of $\vec{r}$ and r, we get
$$\vec{F} = \frac{k(x\hat{i} + y\hat{j} + z\hat{k})}{(x^2+y^2+z^2)^{3/2}},$$
Therefore, the work done by this force in displacement from position $\vec{r}_1 = x_1\hat{i} + y_1\hat{j} + z_1\hat{k}$ to $\vec{r}_2 = x_2\hat{i} + y_2\hat{j} + z_2\hat{k}$ i.e., from (x_1, y_1, z_1) to (x_2, y_2, z_2) is given by
$$W = \int_{(x_1,y_1,z_1)}^{(x_2,y_2,z_2)} \frac{k}{(x^2+y^2+z^2)^{3/2}} [x dx + y dy + z dz]$$
or $W = k\left[\frac{1}{\sqrt{x_1^2+y_1^2+z_1^2}} - \frac{1}{\sqrt{x_2^2+y_2^2+z_2^2}}\right] = k\left(\frac{1}{r_1} - \frac{1}{r_2}\right)$

Here, $r_1 = \sqrt{x_1^2 + y_1^2 + z_1^2}$ and $r_2 = \sqrt{x_2^2 + y_2^2 + z_2^2}$
From above it is clear that, the work done depends only on the initial and final position of the particle not on the path followed by it.

EXAMPLE 34. Prove that $\vec{F} = x\hat{i} + y\hat{j} + z\hat{k}$ is a conservative force. Also find the work done by this force in displacing a particle from (x_1, y_1, z_1) to (x_2, y_2, z_2).

APPROACH Follow the same approach as given in above example.

SOLUTION $\vec{F} = x\hat{i} + y\hat{j} + z\hat{k}$
Work done by above force in motion of particle from position (x_1, y_1, z_1) to (x_2, y_2, z_2) is given by-
$W = \int_{(x_1,y_1,z_1)}^{(x_2,y_2,z_2)} \vec{F} \cdot d\vec{r}$
or $W = \int_{(x_1,y_1,z_1)}^{(x_2,y_2,z_2)} (x\hat{i} + y\hat{j} + z\hat{k}) \cdot (dx\hat{i} + dy\hat{j} + dz\hat{k})$
or $W = \int_{(x_1,y_1,z_1)}^{(x_2,y_2,z_2)} (x dx + y dy + z dz)$
$W = \frac{1}{2}(x_2^2 - x_1^2 + y_2^2 - y_1^2 + z_2^2 - z_1^2)$

which depends only on initial and final positions (x_1, y_1, z_1) and (x_2, y_2, z_2) respectively. Therefore, $\vec{F}$ is a conservative force.

18. POTENTIAL ENERGY

Kinetic energy describes the system's motion, but even a system that has no motion may still have energy associated with the arrangement of the particles in it. This energy, known as **potential energy,** depends on the configuration of a system. Potential energy is only used to describe a system consisting of two or more particles that interact with each other via one or more

---

*Any force, which is directed away or towards a fixed point and depends only on the distance between the center and the particle in the question, is called the central force.

internal forces. That is, forces acting on particle(s) in the system come only from the particle(s) within that system.

The term, 'potential energy' cannot be associated with a single particle because a single (isolated) particle cannot experience internal forces. To associate potential energy with a system, we must define the system so that it includes at least two particles and both the particles interacting with a conservative force or forces.

☞ *Not all types of internal forces can be associated with potential energy. Potential energy can only be associated with internal **conservative** forces of the system.*

Now, if the work done by a conservative force can be expressed as the difference between a quantity $U(x,y,z)$ evaluated at the initial and at the final points. Then, the quantity $U(x,y,z)$ is called the *potential energy* and is a function of the coordinates of the particles.

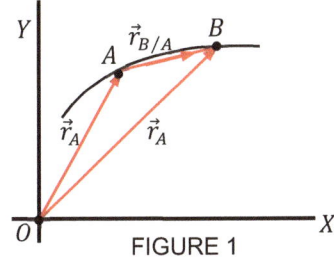

FIGURE 1

If a particle moves from point A to B, then, the work done by system's internal conservative force $\vec{F}$ on this particle,

$$W = \int_A^B \vec{F} \cdot d\vec{r} = U_A - U_B \qquad \ldots (1)$$

Note that we have written $U_A - U_B$ not $U_B - U_A$ that is, the work done is equal to U at the starting point minus U at the endpoint.

Equation (1) can also be written as $U_A - U_B = \int_A^B \vec{F} \cdot d\vec{r}$

or $\qquad U_B - U_A = -\int_A^B \vec{F} \cdot d\vec{r}$

or $\qquad \Delta U = -\int_A^B \vec{F} \cdot d\vec{r}$

Here, $\Delta U = U_B - U_A$.

If $\vec{F}$ is constant, then

$$\Delta U = -\vec{F} \cdot \int_A^B d\vec{r} = -\vec{F} \cdot (\vec{r}_B - \vec{r}_A) \qquad \ldots (2)$$

Strictly speaking, the potential energy U must depend on the coordinates of the particle considered, as well as on the coordinates of all the other particles of the universe which interact with it. But just for simplicity, here, we have assumed that the rest of the universe essentially fixed, and thus only the coordinates of the particle under consideration appear in U.

If you compare Eq. (1) with the work energy theorem, you may realise that the work energy theorem is generally valid no matter what the force $\vec{F}$ may be. It is always true that $K = \frac{1}{2}mv^2$ while the form of the function $U(x,y,z)$ depends on the nature of the force $\vec{F}$, and not all forces may satisfy the condition set by Eq. (1). Only those satisfying it are called *conservative.*

From equation (2), we can define the change in potential energy as follows-
The change in potential energy is defined as the negative of work done by system conservative forces (i.e., internal forces) to change the system configuration.

PURPOSE *By defining PE we can avoid repeated calculation of work for conservative forces and since PE depends only on position (initial and final), we can directly write effect of conservative forces in terms of their respective potential energies.*

☞ If a mass is attached with spring, then we call it as a spring mass system. Now if an external (i.e., a non-conservative) force is pulling or pushing the mass against system's conservative internal force (i.e. spring force) very slowly so that system always remains in equilibrium, then the negative work done by system's internal force (i.e. conservative force) will be equal to change in elastic potential energy of spring mass system.

☞ In case of earth mass system, the gravitational force is an internal force i.e., conservative force applied within the system by given mass and earth on each other. Now if we apply an external force (i.e., a non conservative force) to change the configuration of the earth mass system such that it always remains in equilibrium with system's internal force, then the negative work done by the system's internal force (i.e. conservative force) will be equal to change in gravitational potential energy of the earth mass system.

KEYPOINT In general, potential energy is defined in terms of system's internal forces, not in terms of external forces. So, we should always remember that the change in potential energy of the system is equal to negative of work done by system conservative force.

18.1. GRAVITATIONAL POTENTIAL ENERGY

18.1.1. GRAVITATIONAL POTENTIAL ENERGY WHEN GRAVITATIONAL FORCE IS UNIFORM

Project a stone up with initial speed v_i. Ignoring air resistance, how high does the stone go? Although, we can solve this problem with Newton's second law, but here we use work and energy concept.

The stone's initial kinetic energy is $K_i = \frac{1}{2}mv_i^2$

MECHANICS

For an upward displacement Δy, gravity does negative work $W_{grav} = -mg\,\Delta y$

No other forces act, so this is the total work done on the stone in earth frame. The stone is momentarily at rest at the top, so final kinetic energy, $K_f = 0$.

∴ $\qquad W_{grav} = K_f - K_i$

or $\qquad -mg\Delta y = 0 - \frac{1}{2}mv_i^2$

or $\qquad \Delta y = \frac{v_i^2}{2g}$

From the standpoint of energy conservation, where did the stone's initial kinetic energy go? If total energy cannot change, it must be "stored" somewhere. Furthermore, the stone gets its kinetic energy back as it falls from its highest point to its initial position, so the energy is stored in a way that is easily recovered as kinetic energy. Stored energy due to the interaction of an object with something else (here, Earth's gravitational field) that can easily be recovered as kinetic energy is called **potential energy** (symbol U).

The change in gravitational potential energy when an object moves up or down is the *negative* of the work done by gravity:

Change in gravitational potential energy:

$$\Delta U_{grav} = -W_{grav} \qquad \ldots (1)$$

If the gravitational field is uniform, the work done by gravity, in upward displacement Δy, is-

$W_{grav} = F_y \Delta y = -mg\Delta y$

where the y-axis points up. Therefore,

Change in gravitational potential energy:

$$\Delta U_{grav} = -W_{grav} = mg\Delta y \qquad \ldots (2)$$

(Uniform $\vec{g}$, y-axis up)

Equation (2) holds even if the object does not move in a straight-line path.

SIGNIFICANCE OF THE NEGATIVE SIGN IN EQ. (1).

When the stone moves up, Δy is positive. The gravitational force and the displacement of the stone are in opposite directions, so the work done by gravity is negative, gravity is taking away kinetic energy and adding it to its stored potential energy, so the potential energy increases (Fig. 1a). If the stone moves down, Δy is negative. The work done by gravity is positive; gravity is giving back kinetic energy by depleting its storage of potential energy, so the potential energy decreases (Fig.1b).

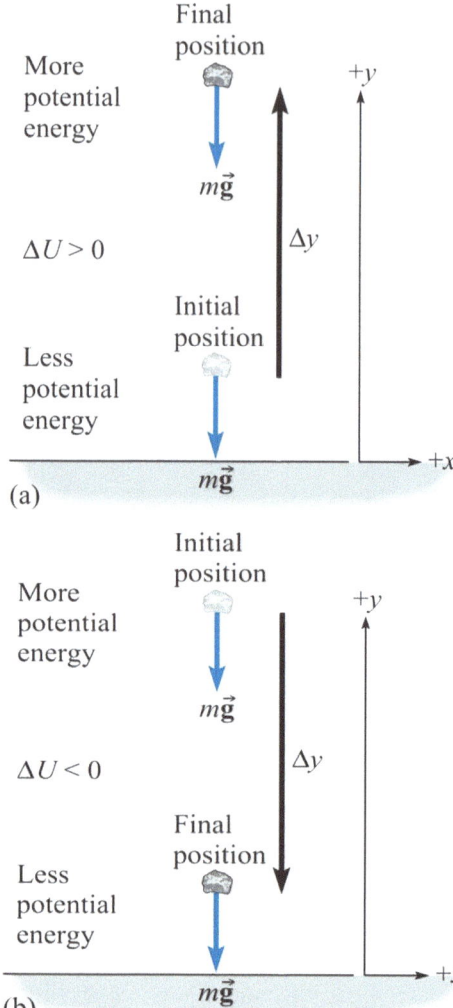

Fig.1 (a) When the stone moves up, the gravitational potential energy increases. (b) When the stone moves down, the gravitational potential energy decreases.

REFERENCE CONFIGURATION

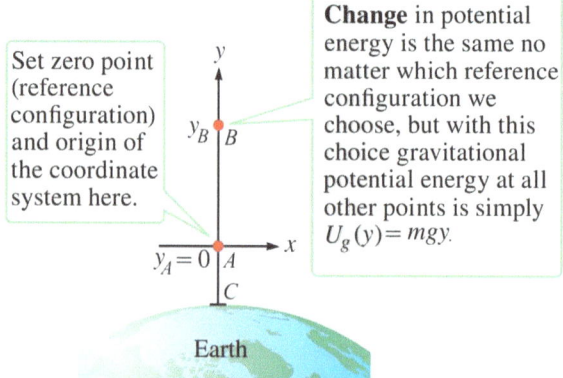

FIGURE 1 Apple Earth system

To make calculations easier, we typically choose a particular configuration to be the *reference configuration* and assign it a potential energy of zero. The potential energy of all other configurations can then be found from a change in potential energy relative

to the reference configuration. Choosing the reference configuration in a situation that involves the Earth and a particle near its surface means setting the origin of the coordinate system to a convenient place. For example, if we place the origin of the coordinate system in Figure 1 at the position A of the apple, it becomes the reference configuration, with a potential energy of zero. Now gravitational potential energy difference between B and A is given by-
$$\Delta U = U_B - U_A = mgy_B - 0 = mgy_B$$
The **gravitational potential energy** of any other configuration may now be written as
$$U_g = mgy \qquad \ldots (1)$$
So, if the apple is at the origin A, the potential energy of the Earth–apple system is zero. Above point A, PE will be positive and below it, PE will be negative. Potential energy is a scalar, so the signs do not indicate direction; rather, they indicate a relative change in potential energy. If the apple is above the origin, the system has more energy than if the apple is below the origin.
If $AC = h$, then $U_C - U_A = U_C - 0 = U_C = -mgh$
Now if we consider ground level (i.e., position C) as a reference level, then
$U_C = 0$, $U_A - U_C = mgh - 0 = mgh$,
here $U_B - U_A = mg(y_B + h) - mgh = mgy_B$
and $\quad U_C - U_A = 0 - mgh = -mgh$

Thus, we can say that potential energy difference always remains same, it doesn't matter where we consider zero potential energy level.
Thus, *by definition we can only find difference of PE not absolute value.*
☞ *Generally, you can choose any level to be the reference level, but once chosen, be consistent.*

18.1.2. GRAVITATIONAL POTENTIAL ENERGY WHEN GRAVITATIONAL FORCE IS NON-UNIFORM

When motion of a body of mass m involves distances from the earth surface large enough, the variation in the gravitational force between the body and the earth cannot be neglected. For such physical situations the configuration, when the body is at infinitely large distance from the earth center is taken as the reference configuration and potential energy of this configuration is arbitrarily assumed zero ($U_\infty = 0$).
If we bring the body very slowly from infinity to a distance r from the earth center, then the work done W_g by the gravitational force is given by the following equation.
$$W_g = \int_\infty^r \vec{F}_g \cdot d\vec{r} = \left[\frac{GMm}{r}\right]_\infty^r$$
Negative of this work done equals to change in potential energy of the system. Denoting potential energies in configuration of separation r and ∞ by U_r and U_∞, we have

$$U_r - U_\infty = -W_g$$
or $\quad U_r = -\frac{GMm}{r} \qquad [\because U_\infty = 0]$

In general, if a particle moves from position $\vec{r}_1$ to $\vec{r}_2$ under the effect of position dependent conservative force $\vec{F}$ and the potential energy of the system changes from U_1 to U_2, then, the change in potential energy is given by-
$$\int_{U_1}^{U_2} dU = -\int_{r_1}^{r_2} \vec{F} \cdot d\vec{r}$$
or $\quad U_2 - U_1 = -\int_{r_1}^{r_2} \vec{F} \cdot d\vec{r}$
or $\quad \Delta U = -\int_{r_1}^{r_2} \vec{F} \cdot d\vec{r}$

EXAMPLE 35. Find the gravitational potential energies in the following physical situations. Assume the ground as the reference potential energy level.

(a) A thin rod of mass m and length L kept at angle θ with one of its end touching the ground.
(b) A flexible rope of mass m and length L placed on a smooth hemisphere of radius R and one of the ends of the rope is fixed at the top of the hemisphere.
APPROACH In both the cases, mass is distributed over a range of position coordinates. Therefore, we first find the potential energy of any elementary part of rod (or rope) and earth system, then integrate it for whole length of the rod (or rope)
SOLUTION (a) If mass density of the rod is assumed to be uniform, then linear mass density $\lambda = \frac{m}{L}$.

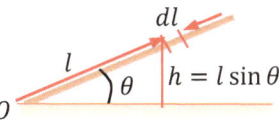

Therefore, the mass of an infinitesimally small portion of length dl of the rod at distance l from the bottom end O is $dm = \lambda dl = \frac{m}{L} dl$
If we consider the reference level of potential energy at ground surface, then the potential energy of this small portion of the rod, $dU = (dm)gh$, here h is the height of the elementary mass dm.
From above figure, $h = l \sin\theta$ and $dm = \frac{m}{L} dl$, therefore-
$$dU = (dm)gh = \left(\frac{m}{L} dl\right) g(l \sin\theta)$$
or $\quad dU = \frac{m}{L} gl \sin\theta \, dl$
To find the total potential energy stored in the rod, we integrate it from $l = 0$ to $l = L$. i.e.,
$$U = \int_0^L \frac{m}{L} gl \sin\theta \, dl = \frac{m}{L} g \sin\theta \left[\frac{l^2}{2}\right]_0^L$$
or $\quad U = \frac{1}{2} mgL \sin\theta$
(b) If mass density of rope is assumed to be uniform, then linear mass density $\lambda = \frac{m}{L}$.

Therefore, the mass of an infinitesimally small portion of length dl of the rope at distance l from the top end $dm = \lambda dl = \frac{m}{L} dl$

Now, if we consider the reference gravitational potential energy level at ground surface, then gravitational potential energy stored in this elementary mass and earth system

$$dU = (dm)gh = \left(\frac{m}{L} dl\right) g(R \sin \theta)$$

$$dU = (dm)gh =$$

From above figure, $dl = R d\theta$

$\therefore \quad dU = \left(\frac{m}{L} R d\theta\right) g(R \sin \theta) = \frac{mg}{L}^2 \sin \theta \, d\theta$

Therefore, total gravitational potential energy stored in rope of length L and earth system,

$U = \frac{mg}{L}^2 \int_0^\alpha \sin \theta \, d\theta$ [see above figure]

or $U = \frac{mg}{L}^2 [-\cos \theta]_0^\alpha = -\frac{mgR^2}{L} [\cos \alpha - \cos 0]$

or $U = \frac{mgR^2}{L} [1 - \cos \alpha]$ [$\because \cos 0 = 1$]

Again, from above figure, $\alpha = \frac{L}{R}$

Therefore, $U = \frac{mg}{L}^2 \left[1 - \cos\left(\frac{L}{R}\right)\right]$

EXAMPLE 36. *A person pushes a block of mass m up a quarter-circle track as shown in Fig. 1 by exerting a variable force. Starting from rest, the mass arrives at the top of the track with a speed of zero. Show that the work done by the force of gravity is path independent and find the change in the block's potential energy.*

APPROACH To calculate the work done by gravity along the track, use relation, $W_g = \int_A^B \vec{F}_g \cdot d\vec{s} = \int_A^B F_g ds \cos \theta$, with $F_g = mg$, $ds = R \, d\beta$ [see Fig. 1b].

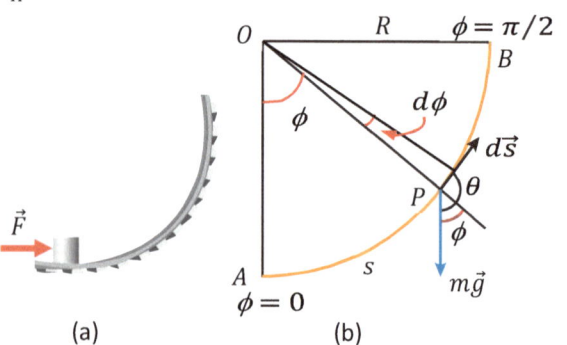

FIGURE 1 (a) A block is pushed along a semicircular track. (b) Diagram showing the force of gravity and the infinitesimal displacement along the track.

Let at any instant, the block is at any general position P of the circular track of radius R. The linear distance of the position P from starting point A is, s. The angular displacement of P from vertical direction is ϕ.

If in infinitesimal displacement vector ds, the angular displacement is $d\phi$, then from Fig. (b),

$$ds = R d\phi. \qquad \ldots (1)$$

Again from Fig. (b), the angle between infinitesimal displacement vector $d\vec{s}$ and downward gravitational force $m\vec{g}$, is-

$$\theta = \phi + \frac{\pi}{2} \qquad \ldots (2)$$

Differentiating both sides, we get

$$d\theta = d\phi$$

Substituting, this value of $d\phi$, from above in equation (1), we get-

$$ds = R d\theta.$$

At position A, $\phi = 0$, therefore,

$$\theta = \phi + \frac{\pi}{2} = \frac{\pi}{2}$$

and at position B, $\phi = \frac{\pi}{2}$, therefore,

$$\theta = \phi + \frac{\pi}{2} = \frac{\pi}{2} + \frac{\pi}{2} = \pi$$

As the block moves along the track, the angle θ between the gravitational force vector and the displacement changes continuously from $\theta = \pi/2$ to $\theta = \pi$.

Therefore, to calculate the work done by gravity along the track, apply the relation

$$dW_g = m\vec{g} \cdot d\vec{s} = (mg)(ds) \cos \theta$$
$$= (mg)(R d\theta) \cos \theta$$

$\therefore \quad W_g = \int_{\pi/2}^{\pi} (mg)(R d\theta) \cos \theta \qquad \ldots (3)$

And the work done by gravity in upward vertical displacement R-

$$W_g' = m\vec{g} \cdot \vec{s} = -mgR \qquad \ldots (4)$$

Now, for first part of the problem, we calculate W_g from (3) and then compare it with work obtained from equation (4).

For second part of the problem, apply the relation,

$$\Delta U = -W_g = -\int_{\pi/2}^{\pi} (mg)(R d\theta) \cos \theta$$

SOLUTION

$W_g = \int_{\pi/2}^{\pi} (mg)(R d\theta) \cos \theta = (mgR) \int_{\pi/2}^{\pi} \cos \theta \, d\theta$
$\qquad = (mgR) \sin \theta |_{\pi/2}^{\pi}$
$\qquad = mgR[\sin \pi - \sin (\pi/2)]$

or $\quad W_g = -mgR \qquad \ldots (5)$

And, the work done by gravity in direct vertical displacement R of the block-

$$W_g' = mgR \cos 180° = -mgR \qquad \ldots (6)$$

From equations, (5) and (4), it is clear that, the work done by gravity remains same whether the block is pushed along the track from point A to B by vertical displacement R, or we take it directly along vertical direction by height R. In other words, we can say that work done by gravity is path independent.

Since, the change in potential energy of the block-Earth system is equal to the negative of the work done by the force of gravity on the block. Therefore,

$$\Delta U = -W_g = mgR$$

☞ this example illustrates the fact that the change in gravitational potential energy is path independent and depends only on the endpoints of the path.

IMPORTANT POINTS

➢ In a sense, potential energy is a storage system for energy. When we increase the separation between an object and the ground, the work we do is stored in the form of an increased potential energy. Not only that, but the storage system is perfect, in the sense that the energy is never lost, as long as the separation remains the same. The object can rest on the shelf for a million years, and still, when it falls, it gains the same amount of kinetic energy.

➢ Work done against friction, however, is not "stored" as potential energy. Instead, it is dissipated into other forms of energy such as heat or sound. The same is true of other nonconservative forces. Only conservative forces have the potential-energy storage system.

➢ **Choose "zero height" to be wherever you like**. When working with gravitational potential energy, we may choose any height to be $y = 0$. If we shift the origin for y, the values of y_1 and y_2 change, as do the values of $U_{grav,1}$ and $U_{grav,2}$. But this shift has no effect on the *difference* in height $y_2 - y_1$ or on the *difference* in gravitational potential energy $U_{grav,2} - U_{grav,1} = mg(y_2 - y_1)$.

18.2. ELASTIC POTENTIAL ENERGY

In case of spring, the natural length position of free end of spring is assumed to be reference point and always assigned zero potential energy (This is a universal assumption). In gravity we can take any point as reference and assign it any value of potential energy.

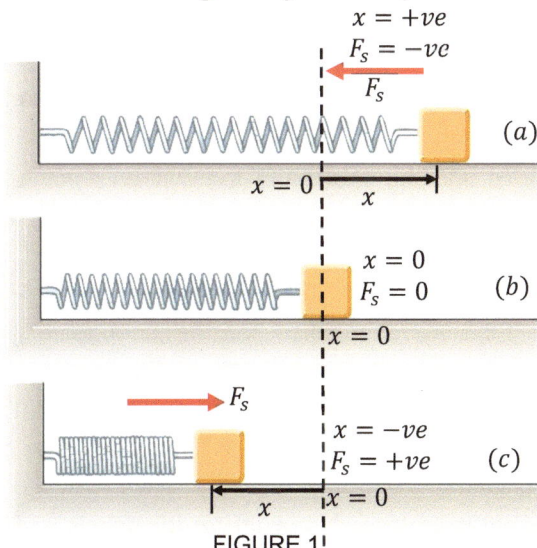

FIGURE 1

Now, considering to and fro motion of a block attached with a spring on a horizontal frictionless table. At any instant the expansion or compression of the spring is denoted by x. At mean position of the block, where the spring is in its natural length, the expansion or compression in the spring, $x = 0$. The reference level for elastic potential energy is always considered at the position, where extension or compression in the spring is zero, i.e., elastic potential energy for spring block system, $U = 0$ for $x = 0$.

For stretching: If at any instant the position of the block from mean position is x and it is moving along $+ve$ direction of x axis, then spring force acting on the block, $\vec{F}_s = -kx\hat{i}$ [Fig.1a]

Here, negative sign shows that, the direction of spring force is opposite to displacement of block.

If we consider an additional displacement dx along positive direction of x axis, then small work done by spring force on the block,

$dW_s = \vec{F}_s \cdot d\vec{s}$, here, $d\vec{s}(= dx\hat{i})$ is the displacement vector.

or $dW_s = (-kx\hat{i}) \cdot (dx\hat{i}) = -kx\,dx$,

∴ Total work done in displacement of the block from initial position $x = 0$ to final position $x = x_m$,

$W = \int_i^f \vec{F}_s \cdot d\vec{s} = \int_0^{x_m} -kx\,dx$

or $W = -k\left[\dfrac{x^2}{2}\right]_0^{x_m} = -\dfrac{1}{2}kx_m^2$

Therefore, change in potential energy of the spring block system,

$U_f - U_i = -\int_i^f \vec{F}_s \cdot d\vec{s} = -\left[-\dfrac{1}{2}kx_m^2\right]$

or $U_f - 0 = \dfrac{1}{2}kx_m^2$

[∵ Elastic potential energy at $x = 0$ is zero, i.e. $U_i = 0$]

or $U = \dfrac{1}{2}kx_m^2$

For compression: When the block is along negative side of x axis, there is a compression in the spring. If at any instant, the compression in the spring is x, then spring force on the block-

$\vec{F}_s = kx\hat{i}$ [Fig.1c]

If block further compresses the spring by an additional distance dx, then displacement vector $d\vec{s} = -dx\hat{i}$

Therefore, change in potential energy in total compression by distance x_m is given by

$U_f - U_i = -\int_i^f \vec{F}_s \cdot d\vec{s} = -\int_0^{x_m}(kx\hat{i})(-dx\hat{i})$

$U = \dfrac{1}{2}kx_m^2$

Thus, if spring is either stretched or compressed from natural length by x the potential energy is $\dfrac{1}{2}kx^2$. Emphasise that for solving problems of spring, always measure distances from natural length.

18.2.1. GRAVITATIONAL POTENTIAL ENERGY VS. ELASTIC POTENTIAL ENERGY.

An important difference between gravitational potential energy $U_{grav} = mgy$ and elastic potential energy $U_{el} = \dfrac{1}{2}kx^2$ is that we cannot choose $x = 0$ to be wherever we wish. In Eq. $U_{el} = \dfrac{1}{2}kx^2$, $x = 0$ must be the position at which the spring is neither stretched nor compressed. At that position, both its elastic potential energy and the force that it exerts are zero.

19. CHECKPOINT 3

1. •A 2.0 kg sloth hangs 5.0 m above the ground
   (a) What is the gravitational potential energy U of the sloth–Earth system if we take the reference point $y = 0$ to be (1) at the ground, (2) at a balcony floor that is 3.0 m above the ground, (3) at the limb, and (4) 1.0 m above the limb? Take the gravitational potential energy to be zero at $y = 0$.
   (b) The sloth drops to the ground. For each choice of reference point, what is the change ΔU in the potential energy of the sloth–Earth system due to the fall?

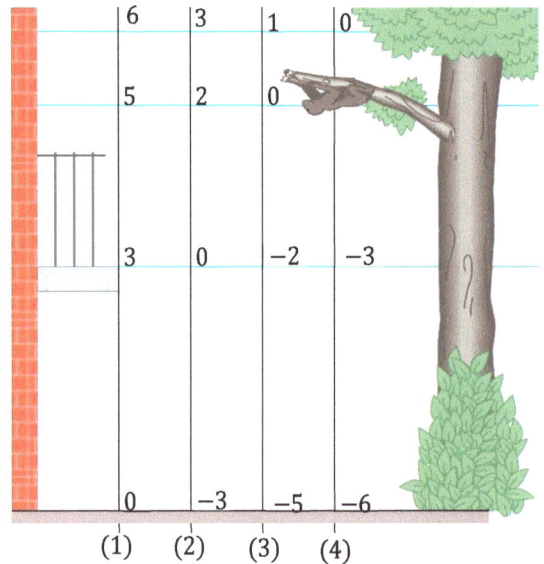

2. •A construction worker is repairing the roof of a house. His hammer of mass 0.55 kg slips from his hand and falls through a basketball hoop to the driveway below. The basketball hoop is 3.0 m above the ground and 5.7 m below the roof where the hammer left the worker's hand.
   (a) To calculate potential energy, what must be in the system?
   (b) Place the origin of an upward-pointing y axis on the driveway. What is the change in the potential energy of the system?
   (c) Now use the basketball hoop as the origin of the y axis and find the change in potential energy.

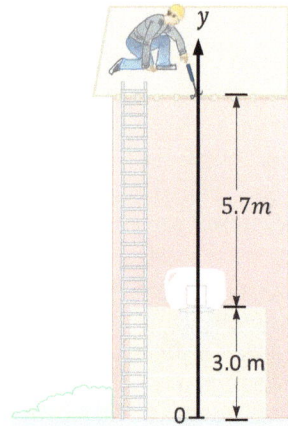

3. ••In an experiment, a ball of mass m, is hanged from a vertical spring of spring constant k (see following figure). The experiment will begin when the ball is displaced straight downward by a distance h.

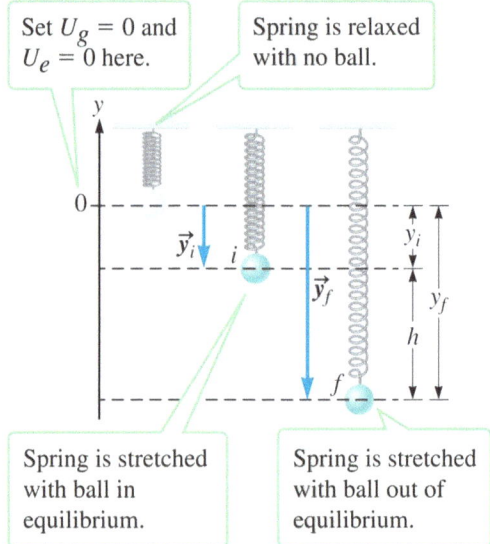

In above figure U_e = elastic PE, U_g = gravitational PE.
(a) To find the change in the system's potential energy, what must be included in the system? (Be sure that there are no external forces exerted on your system.)
(b) Find an expression for the change in the system's potential energy from the equilibrium position (after the ball has been attached to the spring as in above figure to a distance h below equilibrium.

4. •A stone is tossed straight up in the air and is moving upward. (a) Does the gravitational potential energy increase, decrease, or stay the same? (b) What about the kinetic energy? (c) What force, if

any, does work on the stone once it leaves the hand of the one who threw it?

20. MECHANICAL ENERGY

The total work done on an object can always be written as the sum of the work done by conservative forces (W_c) plus the work done by nonconservative forces (W_{nc}). Since the total work is equal to the change in the object's kinetic energy (work energy theorem), therefore we can write

$$W_{tot} = \sum W_C + \sum W_{nc} = K_f - K_i$$

∵ $\sum W_C = -(U_f - U_i)$

∴ $-(U_f - U_i) + \sum W_{nc} = K_f - K_i$

or $\sum W_{nc} = (K_f - K_i) + (U_f - U_i)$

or $\sum W_{nc} = \Delta K + \Delta U$ ... (1)

Thus, the work done by non-conservative forces = change in KE + Change in PE

The sum of the kinetic and potential energies ($KE + PE$) is called the **mechanical energy**, i.e.,

$$K + U = E_{mech} \quad \text{(Mechanical Energy)}$$

Therefore, $\sum W_{nc} = \Delta E_{mech}$,

i.e., W_{nc} is equal to the change in mechanical energy of the system. When finding the change in mechanical energy, do not include the work done by conservative forces. Conservative forces such as gravity do not change the mechanical energy; they just change one form of mechanical energy into another. Work done by conservative forces is already accounted for by the change in potential energy.

☞ When, the net work is done upon an object by an external force, the total mechanical energy (KE + PE) of that object is changed. If the work is positive, then the object will gain energy. If the work is negative, then the object will lose energy. The gain or loss in energy can be in the form of potential energy, kinetic energy, or both. Under such circumstances, the work that is done will be equal to the change in mechanical energy of the object. **Because external forces are capable of changing the total mechanical energy of an object, they are referred to as nonconservative forces**.

➢ If we apply an external force $\vec{F}_{ext}$, against the resultant of internal conservative forces, to change the configuration of system very slowly with zero acceleration, then change in the kinetic energy of the system will be zero, i.e., $\Delta K = K_f - K_i = 0$. In this case, $\sum W_{nc} = \Delta K + \Delta U = 0 + \Delta U = \Delta U$

Therefore, in such a special case we can say that the work done by an external force is equal to the change in potential energy of the system.

Thus, if the configuration of a system changes from A to B, then we can write

$$\Delta U = U_B - U_A = \sum W_{nc} = \vec{F}_{ext} \cdot \int_A^B d\vec{r}$$

$$\Delta U = \vec{F}_{ext} \cdot (\vec{r}_B - \vec{r}_A)$$

Thus, we can also define the change in potential energy as the positive work done by the external force against system's conservative forces, provided that system always remains in dynamic equilibrium i.e., the system configuration changes very slowly with zero acceleration.

> **KEYPOINT** Although the potential energy can only be defined for internal conservative forces of the system but if an external force is always in equilibrium with system conservative force, then we can also write the change in potential energy in terms of external applied force as $\Delta U = \vec{F}_{ext} \cdot (\vec{r}_B - \vec{r}_A)$.

☞ Let us consider a spring block system. If an external (i.e., a non-conservative) force is pulling or pushing the block against system's conservative internal force (i.e. spring force) very slowly so that system always remains in equilibrium, then the work done by external force will be equal to change in elastic potential energy of spring block system.

☞ In case of earth mass system, if we apply an external force (i.e., a non conservative force) to change the configuration of the earth mass system such that it always remains in equilibrium with system's internal force, then the work done by the external non conservative force will be equal to change in gravitational potential energy of the earth mass system.

20.1. POSITIVE VS. NEGATIVE WORK AND ENERGY CHANGE

When work is done by external forces (nonconservative forces), the total mechanical energy of the object is altered. The work that is done can be positive work or negative work depending on whether the force doing the work is directed opposite the object's motion or in the same direction as the object's motion. If the force and the displacement are in the same direction, then positive work is done on the object. If positive work is done on an object by an external force, then the object gains mechanical energy. If the force and the displacement are in the opposite direction, then negative work is done on the object; the object subsequently loses mechanical energy.

21. CONSERVATION OF MECHANICAL ENERGY

When a conservative force does work W_C on a particle, the work-energy theorem tells us that there will be a change in its kinetic energy given by Equation-
$$W_C = \Delta K \qquad \ldots (1)$$
and a change in potential energy given by Equation-
$$W_C = -\Delta U \qquad \ldots (2)$$
From above equations, we can write
$$W_C = \Delta K = -\Delta U$$
$$\Rightarrow \Delta U + \Delta K = \Delta(U + K) = 0 \qquad \ldots (3)$$
Since the sum of the kinetic energy K and potential energy U is called the total mechanical energy of the system, therefore Eq. (3) gives
$$\Delta E_{mech} = 0 \qquad \ldots (4)$$
which is called the **principle of conservation of mechanical energy**.

Thus, when the only forces doing work are conservative forces, energy changes forms- from kinetic to potential (or vice versa); yet the total amount of mechanical is conserved. Note that the conservative forces are the system's internal forces not applied external forces.

☞ When, the only type of force doing net work upon an object is an internal force of system (for example, gravitational and spring forces), the total mechanical energy $(K + U)$ of that object remains constant (or. conserved). This is the reason why the systems internal forces are called conservative forces. In such cases, the object's energy changes form. For example, as an object is "forced" from a high elevation to a lower elevation by gravity, some of the potential energy of that object is transformed into kinetic energy. Yet, the sum of the kinetic and potential energies remains constant.

EXAMPLE 37 Read the description and indicate whether the object gained energy (positive work) or lost energy (negative work)
1) A tee ball player hits a long ball off the tee. During the contact time between ball and bat, the bat is moving at a 10° angle to the horizontal.

2) A person inserts a nail into a block of wood. The hammer head is moving horizontally when it applies force to the nail.

3) The frictional force between highway and tires pushes backwards on the tires of a skidding car.
4) A diver experiences a horizontal reaction force exerted by the blocks upon her feet at start of the race.

5) A weightlifter applies a force to lift a barbell above his head at constant speed.

RESPONSES
1) The force is up and to the right and the displacement is up and to the right. $\vec{F}$ and $\vec{s}$ are in the same direction. Thus, positive work is done.
The applied force of the bat causes the ball to gain both height and speed. Since the force has an up component, it contributes to a height change. Thus, the external or nonconservative force alters the both the kinetic energy and the potential energy of the ball.
2) The force and the displacement both are in downward direction, therefore work done is positive.
The applied force of the hammer causes the nail to gain speed. Thus, the external or nonconservative force alters the kinetic energy of the nail.
3) The force and the displacement are in opposite directions. Thus, negative work is done.
The friction force on the car causes the car to lose speed. Thus, the external or nonconservative force alters the kinetic energy of the car.
4) The force is to the right and the displacement is to the right. $\vec{F}$ and $\vec{s}$ are in the same direction. Thus, positive work is done.
The applied force of the starting blocks causes the diver to gain speed. Thus, the external force alters the kinetic energy of the diver. (NOTE: there is another force - gravity - which changes the diver's height; but the blocks are not responsible for this height change.)
5) The force is up and the displacement is up. $\vec{F}$ and $\vec{s}$ are in the same direction. Thus, positive work is done.

The applied force of the causes the barbell to gain height. Thus, the external or nonconservative force alters the potential energy of the barbell.

22. NON-MECHANICAL ENERGY

Some examples of non-mechanical energies are- sound energy, thermal energy, electrical energy, chemical energy, and nuclear energy. Actually, in very fundamental way every form of energy is either kinetic or potential in nature. Sound energy is contribution of kinetic energy of oscillating molecules and potential energy due intermolecular forces within the medium in which sound propagates. Thermal energy which is a contribution of kinetic energy of disordered motion of molecules in a body and potential energy due to intermolecular forces within the body. Electric energy is kinetic energy of moving charge carries in conductors. Chemical energy is contribution of potential energy due inter-atomic forces. In addition, nuclear energy is contribution of electrostatic potential energy of nucleons.

In fact, every physical phenomenon involves in some way conversion of one form of energy into other. Whenever mechanical energy is converted into other forms or vice versa it always occurs through forces and displacements of their point of applications i.e. work. Therefore, we can say that work is measure of transfer of mechanical energy from one body to other. That is why the unit of energy is usually chosen equal to the unit of work.

23. GENERAL FORM OF CONSERVATION OF ENERGY

In an isolated system, we can relate the total energy at one instant to the total energy at another instant *without considering the energies at intermediate times.*
According to law of conservation of energy, we have
$$\Delta K + \Delta U + \Delta U_{int} = 0$$
here, $\Delta K \rightarrow$ Change in KE,
$\Delta U \rightarrow$ Change in PE,
and $\Delta U_{int} \rightarrow$ Change in internal energy

EXAMPLE 38. *Chain is on the verge of slipping, find the velocity of the chain, when it has slipped.*

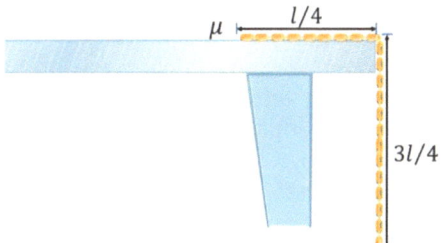

APPROACH Since, we can easily calculate the work done by force of friction on the part of chain placed on the table and the work done by gravity by using the relation $W_c = U_i - U_f$ and the initial kinetic energy of the chain is also known (it is zero), therefore to find the final kinetic energy, apply the principle of the work energy theorem (WRT).

SOLUTION At the verge of slipping, the weight of hanging part of the chain is balanced by the force of friction between the table and the part of chain on the table-
i.e., $f = \frac{3Mg}{4}$

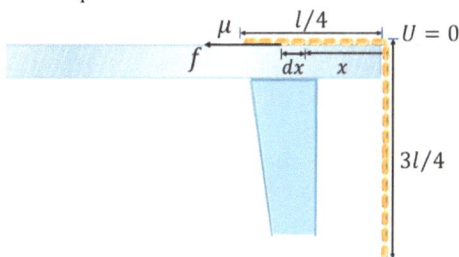

or $\mu \frac{Mg}{4} = \frac{3M}{4} \Rightarrow \mu = 3$

If λ is the mass per unit length of the chain, the mass of the chain of length x will be $x\lambda$ and the force of friction acting on it is μxg (assuming that x is the length of the chain from the edge). So, the work done by friction in shift of dx length of the chain on the table
$$dW = -\left(\mu \frac{M}{L} xg\right)(dx)$$
[Force of friction is opposite to displacement]
Work done by friction force when chain completely slip off the table:
$$W = -\int_{L/4}^{0}\left(\mu \frac{M}{L} xg\right)(dx)$$
$$W = -3\frac{Mg}{L}\left(\frac{x^2}{2}\right)_{L/4}^{0} = -\frac{3MgL}{32}$$
Now, by WET, we have
$$W_{nc} + W_c = K_f - K_i$$
or $W + (U_i - U_f) = K_f - K_i$
or $U_i - U_f = K_f - K_i - W$
or $U_i - U_f = \frac{1}{2}mv^2 - 0 + \frac{3MgL}{32}$
or $U_i - U_f = \frac{1}{2}mv^2 + \frac{3MgL}{32}$
here, $U_i =$ initial PE of the system, $U_f =$ final PE of the system
Let, at the level of table, $U = 0$, then initial gravitational PE of chain-earth system
$$U_i = 0 - \left(\frac{3M}{4}\right)g\frac{3l}{8} = -\frac{9Mg}{32}$$
Final gravitational PE of chain-earth system $= \frac{Mgl}{2}$
$\therefore \left(-\frac{9Mgl}{32}\right) - \left(-\frac{Mgl}{2}\right) = \frac{1}{2}Mv^2 + \frac{3Mgl}{32}$
$\Rightarrow \frac{7Mgl}{32} = \frac{1}{2}Mv^2 + \frac{3Mgl}{32}$
$\Rightarrow \frac{1}{2}Mv^2 = \frac{4Mgl}{32}$
$\Rightarrow v = \frac{1}{2}\sqrt{gl}$

EXAMPLE 39. A chain is held on a frictionless table with $(1/n)$ th part of its length hanging over the edge. If the chain has a length 'L' and a mass 'M'

how much work is required to pull the hanging part back on the table?

APPROACH Since, the table is frictionless, therefore, there is no loss of mechanical energy.

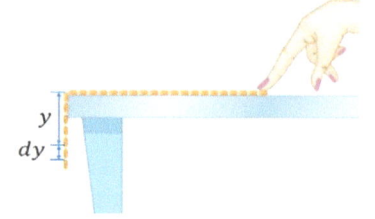

We can find the work done either by using the relation $W_{nc} = U_i - U_f$ or by using the method of work done by a variable force.

SOLUTION **1. By Using the Relation $W_{nc} = \Delta E_{mech}$:**
Let us consider the reference level of gravitational PE at the surface of the table.
According to problem, $\frac{1}{n}$ th part of the chain is hanging over the edge. The mass of this part of the chain is $\frac{M}{n}$. This mass can be considered at the centre of mass of the hanging length i.e., at distance, $\frac{L/n}{2}$ from the edge of the table.
Now, the initial potential energy of the chain earth system-
$$U_i = -\left(\frac{M}{n}\right)g\frac{L/n}{2} = -\frac{MgL}{2n^2}$$
Final potential energy when the hanging part is pulled back on the table
$$U_f = 0$$
∴ work done by non conservative force, i.e., by applied external force, $W_{nc} = \Delta E_{mech} = (U_f - U_i) + (K_f - K_i)$
$$= \left[0 - \left(-\frac{MgL}{2n^2}\right)\right] + (0 - 0) = \frac{MgL}{2n^2}$$

2. Method of Variable Force: If λ is the mass per unit length of the chain, the mass of the chain of length y will be $y\lambda$ and the force acting on it due to gravity will be λyg (assuming that y is the length of the chain hanging over the edge). So, the work done in pulling the dy length of the chain on the table
$$dW = F(-dy) \quad \text{[as } y \text{ is decreasing]}$$
i.e., $\quad dW = (\lambda y g)(-dy) \quad \text{[as } F = \lambda y g\text{]}$
So, the work done in pulling the hanging portion on the table:
$$W = -\int_{L/n}^{0} \lambda y g\, dy = \frac{\lambda g L^2}{2n^2}$$
or $\quad W = \frac{MgL}{2n^2} \quad \text{[as } M = \lambda L\text{]}$

EXAMPLE 40. If a chain of mass M, shown in adjoining figure, starts moving over a massless sprocket, find its KE when chain becomes completely straight.

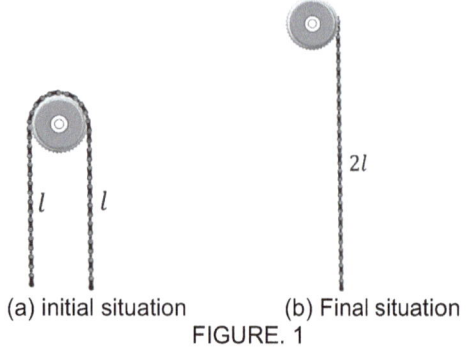

(a) initial situation (b) Final situation
FIGURE. 1

APPROACH Since the force acting on the chain earth system is gravitational force, which is conservative therefore, the mechanical energy of the chain earth system will be conserved. We can easily calculate the initial kinetic energy, initial and final potential energies of the chain, therefore apply the conservation of mechanical energy to chain earth system.

SOLUTION By conservation of mechanical energy, we have
$$K_f + U_f = K_i + U_i \quad \ldots (1)$$
Let us consider the centre of chain sprocket as the reference level for gravitational potential energy.
Initially, the positions of centre of mass of both sides of the chain is at distance, $\frac{l}{2}$ from the centre of chain sprocket, therefore, initial potential energy of the chain earth system-
Since the masses of hanging chain on both sides are $\frac{M}{2}$ for each part and these can be considered at the positions of centre of mass on each side, i.e., at distance $\frac{l}{2}$ from the centre of sprocket.
$$U_i = -\left(\frac{M}{2}g\frac{l}{2}\right) - \left(\frac{M}{2}g\frac{l}{2}\right) = -\frac{Mgl}{2}$$
Initial kinetic energy, $K_i = 0$
When, the chain becomes completely straight, the position of centre of mass of the chain from the centre of the sprocket is at distance l, therefore, the final potential energy of the chain earth system-
$$U_f = -Mgl$$
Substituting these values in (1), we get-
$$K_f + (-Mgl) = 0 + \left(-\frac{Mgl}{2}\right)$$
or $\quad K_f = 0 + \left(-\frac{Mgl}{2}\right) + Mgl = \frac{Mgl}{2}$

EXAMPLE 41. The natural length of the spring shown in figure is $4m$. Find the velocity of ring on the frictionless rod when spring becomes horizontal [Fig.1]. Given that- mass of ring $m = 10\, kg$ and spring constant, $k = 400\, Nm^{-1}$.

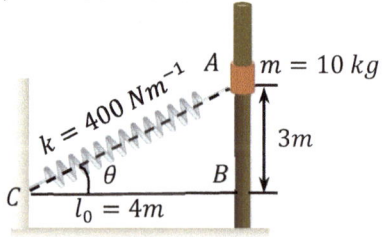

FIGURE 1

APPROACH Since the rod is frictionless, and all forces acting on the ring earth system are conservative; therefore, you can apply the principle of conservation of mechanical energy.

SOLUTION Given that- $m = 10$ kg, $k = 400$ N/m, Natural length of spring $= 4\, m$
Stretched length of spring $CA = \sqrt{(4m)^2 + (3m)^2} = 5m$
Change in length $x = 5m - 4m = 1m$
If we consider the horizontal level of spring as a reference level for gravitational potential energy, then by conservation of mechanical energy, we have-

$$K_i + U_i = K_f + U_f \qquad \ldots (1)$$
here, $K_i = 0$, $K_f = \frac{1}{2}mv^2$, $U_i = \frac{1}{2}kx^2 + mgh$, $U_f = 0$.
Substituting these values in equation (1), we get
$$0 + \left(\frac{1}{2}kx^2 + mgh\right) = \frac{1}{2}mv^2 + 0$$
or $\quad \frac{1}{2} \times 400 \times 1^2 + 10 \times 10 \times 3 = \frac{1}{2} \times 10v^2$
or $\quad 200 + 300 = 5v^2$
or $\quad 5v^2 = 500 \Rightarrow v = 10 m/s$

24. WORK DONE IN MAKING A PYRAMID OF SAND

Such problems can be easily solved in following steps.
Step 1. Assume the net gravitational potential energy of sand particles on the surface of earth as zero
Step 2. Now find the height (h) of center of gravity (CG) of the geometry.
Step 2. Calculate the work done by gravity using relation,
$$W_g = -(U_f - U_i)$$
If external force is always in equilibrium with system's conservative force, then we can write
$$W_{ext} = -W_g = (U_f - U_i)$$
i.e., W_{ext} = change in potential energy.

25. CONSERVATIVE FORCE AS A NEGATIVE GRADIENT OF POTENTIAL ENERGY

25.1. IN ONE DIMENSION

In the one-dimensional case, where a conservative force can be written as a function of x, say, the potential energy can be written as
$$U(x) - U(x_0) = -\int_{x_0}^{x} F_x \cdot dx$$
or $\quad U(x) = -\int_{x_0}^{x} F_x \cdot dx + U(x_0)$
Here $U(x_0)$ is the PE at x_0. It is constant and generally, considered as 0 (reference level PE)
Differentiating both sides with respect to x, we get
$$\frac{dU(x)}{dx} = -F_x$$
Therefore, the conservative force is related to the potential energy function through the relationship-
$$F_x = -\frac{dU(x)}{dx} \qquad \ldots (1)$$
(force from potential energy, one dimension)
That is, the x component of a conservative force acting on a member within a system equals the negative derivative (or slope) of the potential energy of the system with respect to x.
This result makes sense; in regions where $U(x)$ changes most rapidly with x (that is, where $\frac{dU(x)}{dx}$ is large), the greatest amount of work is done during a given displacement, and this corresponds to a large force magnitude. Also, when $F_x(x)$ is in the positive x-direction, $U(x)$ decreases with increasing x. So, $F_x(x)$

and $\frac{dU(x)}{dx}$ should indeed have opposite signs. The physical meaning of Eq. (1) is that *a conservative force always acts to push the system toward lower potential energy.*

KEYPOINT The direction of $\vec{F}$ is always opposite to the direction of increasing potential energy.

As a check, let's consider the function for elastic potential energy, $U(x) = \frac{1}{2}kx^2$.
$$\therefore \quad F_x = -\frac{dU}{dx} = -\frac{d}{dx}\left(\frac{1}{2}kx^2\right) = -kx$$

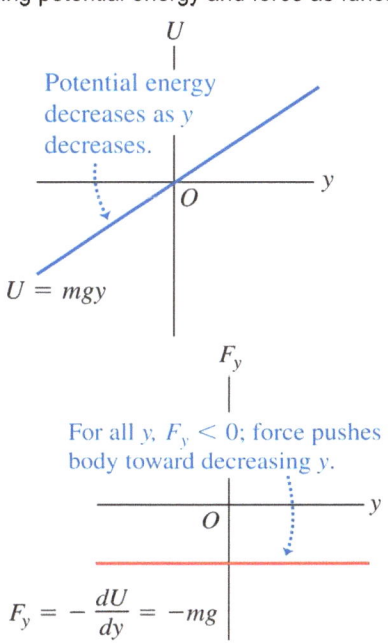

(a) Spring potential energy and force as functions of x

(b) Gravitational potential energy and force as function of y
FIGURE 1. A conservative force is the negative derivative of the corresponding potential energy.

which is the correct expression for the force exerted by an ideal spring (Fig. 1a). Similarly, for gravitational potential energy, we have
$$U(y) = mgy$$
$$\therefore F_y = -\frac{dU}{dy} = -\frac{d}{dy}(mgy) = -mg$$
which is the correct expression for gravitational force.

25.2. IN THREE DIMENSIONS

In three dimensions the force is defined as the negative of the gradient of the potential-energy function i.e.,
$$\vec{F} = -\vec{\nabla}U \qquad \ldots (2)$$
here, $\vec{\nabla}U$ is called the gradient of U.
Since, $\vec{\nabla} = \frac{\partial}{\partial x}\hat{\imath} + \frac{\partial}{\partial y}\hat{\jmath} + \frac{\partial}{\partial z}\hat{k}$
$$\therefore \vec{F} = -\vec{\nabla}U = -\left(\frac{\partial}{\partial x}\hat{\imath} + \frac{\partial}{\partial y}\hat{\jmath} + \frac{\partial}{\partial z}\hat{k}\right)U$$
$$= -\left(\frac{\partial U}{\partial x}\hat{\imath} + \frac{\partial U}{\partial y}\hat{\jmath} + \frac{\partial U}{\partial z}\hat{k}\right)$$
$$\Rightarrow F_x\hat{\imath} + F_y\hat{\jmath} + F_z\hat{k} = -\left(\frac{\partial U}{\partial x}\hat{\imath} + \frac{\partial U}{\partial y}\hat{\jmath} + \frac{\partial U}{\partial z}\hat{k}\right)$$
Comparing both sides, we get
$$F_x = -\frac{\partial U}{\partial x},\ F_y = -\frac{\partial U}{\partial y}\ \text{and}\ F_z = -\frac{\partial U}{\partial z} \qquad \ldots (3)$$
As a check, let's substitute into Eq. (2) the function $U = mgy$ for gravitational potential energy
$$\vec{F} = -\vec{\nabla}(mgy) = -\left(\frac{\partial(mgy)}{\partial x}\hat{\imath} + \frac{\partial(mgy)}{\partial y}\hat{\jmath} + \frac{\partial(mgy)}{\partial z}\hat{k}\right)$$
or $\qquad \vec{F} = (-mg)\hat{\jmath} \qquad \ldots (4)$
This is just the familiar expression for the gravitational force.

EXAMPLE 42. Find the conservative force corresponding to potential energy function $U = 4x^2y + 2yz^2$

APPROACH In three dimensions, the x, y and z components of conservative force, $\vec{F} = F_x\hat{\imath} + F_y\hat{\jmath} + F_z\hat{k}$, are given as-
$$F_x = -\frac{\partial U}{\partial x},\ F_y = -\frac{\partial U}{\partial y}\ \text{and}\ F_z = -\frac{\partial U}{\partial z}$$
On substituting the values of F_x, F_y and F_z in expression for force, you will get the required force.

SOLUTION $U = 4x^2y + 2yz^2$
$$F_x = -\frac{\partial U}{\partial x} = -8xy,\ F_y = -\frac{\partial U}{\partial y} = -(4x^2 + 2z^2),$$
$$F_z = -\frac{\partial U}{\partial z} = -4yz$$
$$\vec{F} = F_x\hat{\imath} + F_y\hat{\jmath} + F_z\hat{k}$$
or $\qquad \vec{F} = -8xy\hat{\imath} - (4x^2 + 2z^2)\hat{\jmath} + -4yz\hat{k}$

EXAMPLE 43. Find the conservative force corresponding to PE function $U = 4r^3$.
APPROACH If U depends on only one variable lets say r, then $\vec{F} = -\left(\frac{dU}{dr}\right)\hat{r}$
SOLUTION Given that-
$$U = 4r^3$$
$$\therefore \vec{F} = -\left(\frac{dU}{dr}\right)\hat{r} = -\left(\frac{d}{dr}4r^3\right)\hat{r}$$
or $\qquad \vec{F} = -(12r^2)\hat{r}$
or $\qquad \vec{F} = 12r^2(-\hat{r})$

EXAMPLE 44. Force between the atoms of a diatomic molecule has its origin in the interactions between the electrons and the nuclei present in each atom. This force is conservative and associated potential energy $U(r)$ is, to a good approximation, represented by the Lennard – Jones potential
$$U(r) = U_o\left\{\left(\frac{a}{r}\right)^{12} - \left(\frac{a}{r}\right)^6\right\}$$
Here r is the distance between the two atoms and U_0 and a are positive constants. Develop expression for the associated force and find the equilibrium separation between the atoms.

APPROACH If U depends on only one variable lets say r, then $\vec{F} = -\left(\frac{dU}{dr}\right)\hat{r}$
or $\qquad F = -\frac{dU}{dr}$
At equilibrium state, the force $F = 0$. Now, solve for r.

SOLUTION Using equation $F = -\frac{dU}{dr}$, we obtain the expression for the force
$$F = \frac{6U_0}{a}\left\{2\left(\frac{a}{r}\right)^{13} - \left(\frac{a}{r}\right)^7\right\}$$
At equilibrium the force must be zero. Therefore, the equilibrium separation r_0 is
$$r_0 = 2^{1/6}a$$

26. ENERGY DIAGRAMS

26.1. THE POTENTIAL ENERGY CURVE

If only conservative forces do work on an object, we can get a lot of insight into its possible motion simply by examining a potential energy diagram - the graph of $U(x)$ versus x. Figure 1 shows a hypothetical but more general potential-energy function $U(x)$. The total energy $E = K + U$ is constant and can be represented as a horizontal line on this graph. Four different possible values for E are shown, labeled E_0, E_1, E_2, and E_3. What the actual value of E will be for a given system depends on the initial conditions. (For example, the total energy E of a mass oscillating at the end of a spring depends on the amount the spring is initially compressed or stretched.) Kinetic energy $K = \frac{1}{2}mv^2$ cannot be less than zero (v would be imaginary), and because $E = U + K = $ constant, therefore, $U(x)$ must be less than or equal to E for all situations: $U(x) \leq E$. Thus the minimum value which the total energy can take for the potential energy shown in Fig. 1 is that labelled E_0. For this value of E, the mass can only be at rest at $x = x_0$. The system has potential energy but no kinetic energy at this position.

WORK, ENERGY AND POWER

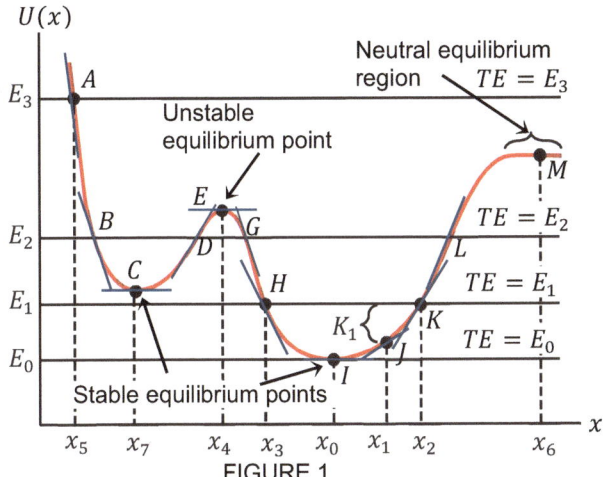

FIGURE 1

If the system's total energy E is greater than E_0, say it is E_1 on our plot, the system can have both kinetic and potential energy. Because energy is conserved,
$$K = E - U(x)$$
since the curve represents $U(x)$ at each x, the kinetic energy at any value of x is represented by the distance between the E line and the curve $U(x)$ at that value of x. In the diagram, the kinetic energy for an object at x_1, when its total energy is E_1, is indicated by the notation K_1

An object with energy E_1 can oscillate only between the points x_2 and x_3. This is because if $x > x_2$ or $x < x_3$, the potential energy would be greater than E meaning $K = \frac{1}{2}mv^2 < 0$ and v would be imaginary, and so impossible. At x_2 and x_3 the velocity is zero, since $E = U$ at these points. Hence x_2 and x_3 are called the *turning points* of the motion. These points are indicated by K and H respectively on the U-x graph If the object is at x_0, say, moving to the right, its kinetic energy (and speed) decreases until it reaches zero at $x = x_2$. The object then reverses direction, proceeding to the left and increasing in speed until it passes x_0 again. It continues to move, decreasing in speed until it reaches $x = x_3$ where again $v = 0$, and the object again reverses direction.

If the object has energy $E = E_2$ in Fig. 1, there are four **turning points** B, D, G and L respectively. The object can move in only one of the two potential energy "valleys," BCD or GIL, depending on where it is initially. It cannot get from one valley to the other because of the barrier between them — for example at a point such as x_4 (point E on U-x graph), $U > E_2$, which means v would be imaginary[†].

For energy E_3, there is only one turning point A, at position x_5, since $U(x) < E_3$ for all $x > x_5$. Thus our object, if moving initially to the left, varies in speed as it passes the potential valleys but eventually stops and

turns around at $x = x_5$. It then proceeds to the right indefinitely, never to return.

How do we know the object reverses direction at the turning points? Because of the force exerted on it. The force F is related to the potential energy U by equation, $F = -dU/dx$. The force F is equal to the negative of the slope of the U-versus-x curve at any point x. At $x = x_2$, for example, the slope is positive (see the tangent at point K on the $U - x$ graph) so the force is negative, which means it acts to the left (toward decreasing values of x).

Following Table shows the turning points for different values of total energies (TE) of the particle.

TE ($U = TE$)	Turning Points
E_1	H, K (two turning points)
E_2	B, D, G and L (four turning points)
E_3	A (only one turning point)

At $x = x_0$ the slope is zero, so $F = 0$. At such a point the particle is said to be in equilibrium. This term means simply that the net force on the object is zero. Hence, its acceleration is zero, and so if it is initially at rest, it remains at rest. If the object at rest at $x = x_0$ were moved slightly to the left or right, a nonzero force would act on it in the direction to move it back toward x_0. An object that returns toward its equilibrium point when displaced slightly is said to be at a point of **stable equilibrium**. Any *minimum* in the potential energy curve represents a point of stable equilibrium.

An object at $x = x_4$ would also be in equilibrium, since $F = -dU/dx = 0$ If the object were displaced a bit to either side of x_4, a force would act to pull the object away from the equilibrium point. Points like x_4, where the potential energy curve has a maximum, are points of unstable equilibrium. The object will not return to equilibrium if displaced slightly, but instead will move farther away.

At different points on the U-x graph the positive or negative signs of slope and direction of conservative force acting on the particle is shown in following table-

Points	slope	Direction of Force $\left[F = -\dfrac{dU}{dx} = -slope\right]$
A	$-ve$	$+ve$ direction of x axis
B	$-ve$	$+ve$ direction of x axis
C	0	0
D	$+ve$	$-ve$ direction of x axis
E	0	0
G	$-ve$	$+ve$ direction of x axis
H	$-ve$	$+ve$ direction of x axis
I	0	0
J	$+ve$	$-ve$ direction of x axis
K	$+ve$	$-ve$ direction of x axis
L	$+ve$	$-ve$ direction of x axis

[†] Although this is true according to Newtonian physics, modern quantum mechanics predicts that objects can "tunnel" through such a barrier, and such processes have been observed at the atomic and subatomic level.

When an object is in a region over which U is constant, such as near $x = x_6$ in Fig. 1 the force is zero over some distance $\left[\because F = -\frac{dU}{dx} = -\frac{d}{dx}(\text{constant}) = 0\right]$. The object is in equilibrium and if displaced slightly to one side the force is still zero. The object is said to be in *neutral equilibrium* in this region.

TACTICS BOX

1. The distance from the axis to the PE (potential energy) curve is the particle's potential energy. The distance from the PE curve to the TE (total energy) line is its kinetic energy. These are transformed as the position changes, causing the particle to speed up or slow down, but the sum $K + U$ doesn't change.
2. A point where the TE line crosses the PE curve is a turning point. The particle reverses direction at turning point.
3. The particle cannot be at a point where the PE curve is above the TE line.
4. The PE curve is determined by the properties of the system—mass, spring constant, gravitational constant etc. You cannot change the PE curve. However, you can raise or lower the TE line simply by changing the initial conditions to give the particle more or less total energy.
5. A minimum in the PE curve is a point of stable equilibrium. A maximum in the PE curve is a point of unstable equilibrium.
6. The direction of conservative force acting on the particle at any point on the U-x graph is determined by slope of tangent at that point. The conservative force acting on the particle at any point-
$$F = -\frac{dU}{dx} = -\text{ slope of tangent at that point}$$
7. At equilibrium points, the conservative force on the particle, $F = -\frac{dU}{dx} = 0$, For a stable equilibrium, the direction of conservative force on the particle on either side of equilibrium point is towards the equilibrium point and for an unstable equilibrium point, the direction of conservative force on the particle on either side of equilibrium is always away from equilibrium point.
8. At **stable equilibrium points**, perturbations (small changes in position around the equilibrium position) result in small oscillations around the equilibrium point. At unstable equilibrium points, perturbations result in an accelerated movement of the particle away from the equilibrium point.
9. From mathematical stand point-
$\frac{dU}{dx} = 0, \frac{d^2U}{dx^2} < 0 \Rightarrow$ minima, stable equilibrium
$\frac{dU}{dx} = 0, \frac{d^2U}{dx^2} > 0 \Rightarrow$ maxima, unstable equilibrium
$\frac{dU}{dx} = 0, \frac{d^2U}{dx^2} = 0 \Rightarrow$ neutral equilibrium
10. **Turning points** are points where the kinetic energy is zero and net force reverses the direction of motion of the particle.
11. At neutral equilibrium region, $U =$ constant.

☞ If for small disturbances, the ball returns to its original state, and for larger disturbances, it crosses the hill. then the equilibrium is called **conditionally stable** or **metastable**.

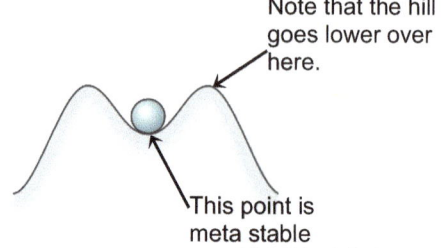

Note that the hill goes lower over here.

This point is meta stable

This block is in metastable equilibrium; it takes a little energy to tip it on edge, but then it would fall over.

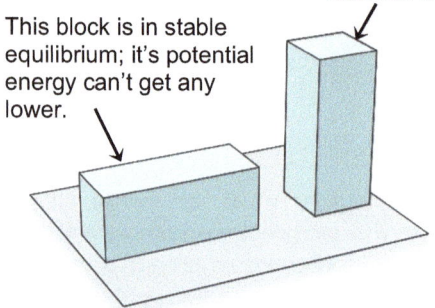

This block is in stable equilibrium; it's potential energy can't get any lower.

FIGURE 2. Identical blocks in stable and metastable equilibria.

EXAMPLE 45. *The potential energy of a system of two particles is given by $U(x) = \frac{a}{x^2} - \frac{b}{x}$. Find the minimum potential energy of the system, where x is the distance of separation; a, b are positive constants.*

APPROACH From mathematical stand point-
$F = -\frac{dU}{dx} = 0, \frac{d^2U}{dx^2} < 0 \Rightarrow$ minima, stable equilibrium
So, find the value of x, corresponding to $F = -\frac{dU}{dx} = 0$ and show that, $\frac{d^2U}{dx^2} < 0$. Now, on substituting this value of x, in the given expression for $U(x)$, you get the minimum value of U.

SOLUTION The given potential energy function is
$$U(x) = \frac{a}{x^2} - \frac{b}{x} \qquad \ldots (1)$$
Differentiating both sides of above equation with respect to x, we get
$$\frac{dU(x)}{dx} = \frac{d}{dx}\left[\frac{a}{x^2} - \frac{b}{x}\right] = \frac{-2a}{x^3} + \frac{b}{x^2} \qquad \ldots (2)$$
$$\therefore \quad F = -\frac{dU(x)}{dx} = -\left[\frac{-2a}{x^3} + \frac{b}{x^2}\right]$$
When the particle is in equilibrium
$$F = 0$$
or $\quad \frac{2a}{x^3} - \frac{b}{x^2} = 0$
or $\quad \frac{1}{x^2}\left[\frac{2a}{x} - b\right] = 0$
$\because x \neq \infty, \therefore \frac{2a}{x} - b = 0 \quad$ or $\quad x = \frac{2a}{b}$
Again differentiating Eq. (2), with respect to x, we get-
$$\frac{d^2U}{dx^2} = \frac{6a}{x^4} - \frac{2b}{x^3} \qquad \ldots (3)$$
Substituting, the value of x, in (3), we get

$$\frac{d^2U}{dx^2} = \frac{6a}{(2a/b)^4} - \frac{2b}{(2a/b)^3} = \frac{6ab^4}{16a^4} - \frac{2b^4}{8a^3}$$

or $\frac{d^2U}{dx^2} = \frac{b^4}{4a^3}\left(\frac{3}{2} - 1\right)$

Since, a, b are positive constants, therefore, $\frac{d^2U}{dx^2} > 0$ which means, there is a minima at $x = \frac{2a}{b}$.

Therefore, the minimum potential energy of the system is obtained by putting $x = \frac{2a}{b}$ in $U(x) = \frac{a}{x^2} - \frac{b}{x}$,

$U_{min} = \frac{a}{\left(\frac{2a}{b}\right)^2} - \frac{b}{\frac{2a}{b}} = \frac{ab^2}{4a^2} - \frac{b^2}{2a} = \frac{b^2}{2a}\left[\frac{1}{2} - 1\right] = -\frac{b^2}{4a}$.

27. POWER

Power is defined as the *rate at which work is done*. The *average power*, equals the work ΔW done divided by the time Δt it takes to do it:

$$\bar{P} = \frac{\Delta W}{\Delta t} = \frac{\text{total work}}{\text{total time}}$$

When energy is converted from one form to another within a system, the average power for the conversion is found in a similar way. For example, if we are interested in finding the average power dissipated by friction, we replace ΔW by the change in thermal energy ΔE:

$$\bar{P} = \frac{\Delta E}{\Delta t} = \frac{\text{total energy tranformed}}{\text{total time}}$$

Average Power (rate at which work is done):

$$\bar{P} = \frac{\Delta W}{\Delta t} = \frac{\Delta E}{\Delta t} \qquad \ldots (1)$$

Keeping with our velocity analogy, just as instantaneous velocity is simply called *velocity* and is given by the time derivative of position, so instantaneous power is simply called *power* and is given by the time derivative of energy transferred or converted. So, for power transferred to or from a system through work, the instantaneous power is

$$P = \lim_{\Delta t \to 0} \frac{\Delta W}{\Delta t} = \frac{dW}{dt} = \frac{\vec{F}.d\vec{x}}{dt},$$

$[\because dW = \vec{F}.d\vec{x}]$

where $\vec{F}$ is the force and $d\vec{x}$ is the displacement in time dt.

$$P = \frac{\vec{F}.d\vec{x}}{dt} = \vec{F}.\vec{v} = Fv\cos\theta$$

($\because \vec{v} = \frac{d\vec{x}}{dt}$ = the instantaneous velocity of the body)

$P = (F\cos\theta) v$
 = (component of $\vec{F}$ along $\vec{v}$)
  × magnitude of velocity

[here θ is the angle between $\vec{F}$ and $\vec{v}$]

If we are interested in finding the power in converting from one form of energy to another, we take the time derivative of that energy converted. Again, for the conversion of mechanical energy to thermal energy, the power dissipated is

$$P = \frac{dE}{dt}$$

Instantaneous power (rate at which work is done):

$$P = \frac{dE}{dt} = \frac{dW}{dt} = \frac{\vec{F}.d\vec{x}}{dt} = \vec{F}.\vec{v} = Fv\cos\theta \qquad \ldots (2)$$

(here $\vec{v}$ is the velocity of the point at which the force $\vec{F}$ is applied, θ is the angle between $\vec{F}$ and $\vec{v}$)

Again, by work energy theorem, we have
$dW = dK$

$\therefore \quad P = \frac{dW}{dt} = \frac{dK}{dt}$

If we need to calculate average power then

$$P_{av} = \frac{\text{Net work done}}{\text{Time taken}} = \frac{\int_0^T P.dt}{T}$$

Average power in time range t_1 to t_2 is given by-

$$P_{av} = \frac{\int_{t_1}^{t_2} P dt}{\int_{t_1}^{t_2} dt}$$

If power is given as a function of velocity, then its average value in velocity range v_1 to v_2 is given by-

$$P_{av} = \frac{\int_{v_1}^{v_2} P dv}{\int_{v_1}^{v_2} dv}$$

> **KEYPOINT** Average value of any general function of x, for example $f(x)$ in the range x_1 to x_2 is defined as- $f_{av} = \frac{\int_{x_1}^{x_2} f(x) dx}{\int_{x_1}^{x_2} dx}$

EXAMPLE 46. If a particle is moving on straight line and a constant instantaneous power is supplying to the particle then find displacement of particle as a function of time.

APPROACH Given that, $P = \vec{F}.\vec{v}$ = constant

First, substitute force $F = mv\frac{dv}{dx}$ and integrate for v as a function of x. Now, put $v = \frac{dx}{dt}$, in the final expression of v and solve for x as a function of t.

SOLUTION Given that power is constant, i.e.,
$P = \vec{F}.\vec{v}$ = constant

Since, $\vec{F}$ and $\vec{v}$ both are along the same straight line, therefore, we can write-

$P = Fv = c$, constant

$\Rightarrow mav = c$

$\because \quad a = v\frac{dv}{dx}$

$\therefore \quad m\left(v\frac{dv}{dx}\right)v = c$

or $v^2 dv = \frac{c}{m} dx$

$\Rightarrow \int_0^v v^2 dv = \int_0^x \frac{c}{m} dx$

$\Rightarrow \frac{v^3}{3} = \frac{c}{m} x \Rightarrow v = \left(\frac{3c}{m} x\right)^{1/3}$

$\Rightarrow \frac{dx}{dt} = \left(\frac{3c}{m} x\right)^{1/3} \Rightarrow \int_0^x \frac{dx}{x^{1/3}} = \int_0^t \left(\frac{3c}{m}\right)^{1/3}$

$\Rightarrow x = \left[\frac{2}{3}\left(\frac{3c}{m}\right)^{1/3} t\right]^{3/2}$

EXAMPLE 47. A small body of mass m is located on a horizontal plane. The body acquires a horizontal velocity v_0. Find the mean power developed by the engine against the frictional force, during the whole time of its motion. Given that- the coefficient of kinetic friction $\mu_k = 0.27$, mass of the body $m = 1kg$, and $v_0 = 1.5\ m/s$.

APPROACH If v is the instantaneous velocity of the given body, then instantaneous power consumed by force of friction,

$$P_{friction} = -\mu_k Nv = -\mu_k mgv \qquad \ldots (1)$$

Here, we have used $-ve$ sign because the direction of kinetic friction is opposite to the velocity v.

Therefore, the power produced by engine against friction,

$$P = -P_{friction} = \mu_k Nv = \mu_k mgv \qquad \ldots (2)$$

Equation (2), represents the power generated against the friction as a function of velocity v.

Now, you can calculate the average power in the velocity range v_1 to v_2 by using the expression-

$$P_{av} = \frac{\int_{v_1}^{v_2} P\,dv}{\int_{v_1}^{v_2} dv}$$

SOLUTION Power produced by engine against the friction

$$P = \mu_k Nv = \mu_k mgv$$

Therefore, the average power in the velocity range 0 to v_0 is given by

$$P_{av} = \frac{\int_0^{v_0} P\,dv}{\int_0^{v_0} dv} = \frac{\mu_k mg}{v_0}\int_0^{v_0} v.dv = \frac{\mu_k mg v_0}{2}$$

Substituting the given values, we get

$$P_{av} = \frac{\mu_k mg\ 0}{2} = \frac{0.27\times 1\times 9.8\times 1.5}{2} = 1.98\ watt$$

EXAMPLE 48. A block of mass $0.5\ kg$ is kept on a rough inclined plane making an angle of 30° with horizontal. What power will be required to move the block up the plane (along the plane) with a velocity of $5m/s$? (Take $\mu = 0.2$ between block and plane)

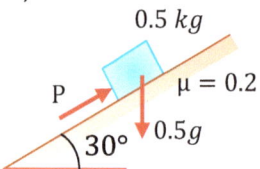

APPROACH Since the block has to move up the incline with constant velocity $5\ m/s$, therefore, this is the case of dynamic equilibrium. In this case, net upward force along the plane F = net downward force along the plane.

Now, the power is given by-

$$P = \vec{F}.\vec{v} = Fv\cos 0 = Fv$$

SOLUTION In dynamic equilibrium,
Net upward force along the plane F = net downward force along the plane.
i.e., $F = 0.5g\sin 30° + \mu(0.5g\cos 30°)$

$$= 0.5\times 10\left(\frac{1}{2} + 0.2\times\frac{\sqrt{3}}{2}\right)$$

$$= 5[0.5 + 0.173] = 3.365\ N.$$

Now the power required to move up along the inclined power $P = Fv$

$$= 3.365\times 5 = 16.825\ N.m/s$$

EXAMPLE 49. A vehicle of mass m starts moving such that its speed v varies with distance traveled s according to the law $v = k\sqrt{s}$, where k is a positive constant. Deduce a relation to express the instantaneous power delivered by its engine.

APPROACH If a particle travels along a curvilinear path, then it will have tangential acceleration $\left(a_T = v\frac{dv}{ds}\right)$ as well as normal acceleration $\left(a_N = \frac{v^2}{R}\right)$, here R is the radius of curvature at the position of particle.

Now, if corresponding to these accelerations, the tangential and normal forces on the particle are, $\vec{F}_T$ and $\vec{F}_N$ respectively, then Net force on the particle, $\vec{F} = \vec{F}_T + \vec{F}_N$ (Fig.1)

Now, the instantaneous power delivered by the engine can be calculated by-

$$P = \vec{F}\cdot\vec{v} = (\vec{F}_T + \vec{F}_N).\vec{v}$$

Now, solve for P.

SOLUTION Let the particle is moving on a curvilinear path. When it has travelled a distance s, the force F acting on it and its speed v are shown in the adjoining figure.

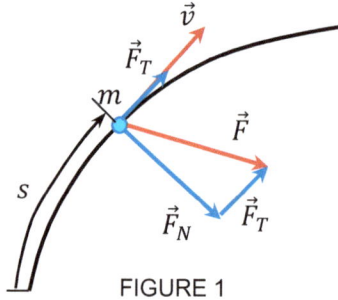

FIGURE 1

Instantaneous power delivered by the engine:
$$P = \vec{F}\cdot\vec{v} = (\vec{F}_T + \vec{F}_N).\vec{v} = F_T v = ma_T v$$
$$[\because \quad \vec{F}_N \perp \vec{v}, \therefore \vec{F}_N.\vec{v} = 0]$$

Tangential acceleration of the vehicle: $a_T = v\frac{dv}{ds}$

$$\therefore \quad P = ma_T v = m\left(v\frac{dv}{ds}\right)v = mv^2\frac{dv}{ds}$$

$$= mv^2\frac{d}{ds}(k\sqrt{s}) = mv^2\left(\frac{k}{2\sqrt{s}}\right)$$

$$= m(k\sqrt{s})^2\left(\frac{k}{2\sqrt{s}}\right) = \frac{mk^3}{2}\sqrt{s}$$

27.1. PHYSICAL SIGNIFICANCE OF POWER

If you want to climb a twelve-story building, you have to do work with respect to the ground which will be equal to your weight ($= mg$ say) times the height of the building. you can climb the building in one minute; you can climb the building in one hour. The work done will be the same, but the power is a measure of how rapidly

work is done. That's why it's the product of force and velocity or work divided by time.

27.2. MEANING OF A 60 W BULB

If you have a $60\,W$ bulb, it's consuming energy at the rate of 60 joules per second.

28. EFFICIENCY

An important characteristic of all engines is their overall efficiency η, defined as the ratio of the useful power output of the engine, P_{out} to the power input, P_{in};
$$\eta = \frac{P_{out}}{P_{in}}$$
The efficiency is always less than 1.0 because no engine can create energy, and in fact, cannot even transform energy from one form to another without some fraction of it going to friction, thermal energy, and other nonuseful forms of energy. For example, an automobile engine converts chemical energy released in the burning of gasoline into mechanical energy that moves the pistons and eventually the wheels. But nearly 85% of the input energy is "wasted" as thermal energy that goes into the cooling system or out the exhaust pipe, plus friction in the moving parts. Thus, car engines are roughly only about 15% efficient.

Power is a scalar with the dimensions of energy per time. Its SI units are joules per second or newton meters per second. This combination of units is called the *watt*, (W), named for the English inventor James Watt. One watt equals 1 joule per second: $1\,W = 1\,J/s$. The kilowatt ($1kW = 10^3 W$) and the megawatt ($1\,MW = 10^6\,W$) are also commonly used.
Another common unit of power is the *horsepower* (hp)
$$1\,hp = 746\,W = 0.746\,kW$$
The *kilowatt-hour* $(kW.h)$ is the usual commercial unit of electrical energy. One kilowatt-hour is the total work done in 1 hour ($3600\,s$) when the power is 1 kilowatt ($10^3\,J/s$), so
$$1kW.h = (10^3 J/s)(3600\,s) = 3.6 \times 10^6 J = 3.6\,MJ$$

> **KEYPOINT** The kilowatt-hour is a unit of *work* or *energy,* not power.

29. SOME IMPORTANT CONVERSION FACTORS

$1\,joule = 1\,newton \times 1\,m = 10^5\,dyne \times 10^2\,cm$
$\qquad = 10^7\,erg$
$1\,watt = 1\,Joule/sec = 10^7\,erg/sec.$
$1\,ft.\,poundal = 1\,poundal \times 1\,ft$
$\qquad = 13825\,dyne \times 30.48\,cm = 4.214 \times 10^5\,erg$
$1kg.wt \times 1\,m = g\,newton \times 1m = g\,joule = 9.8\,joule$
$\quad 1ft.\,lb = g\,poundal \times 1\,ft = 32.2\,poundal \times 1ft$
$\qquad = 32.2\,ft.\,poundal$
$1\,kwh = 10^3\,watt \times 1\,hr = 10^3\,watt \times 3600\,sec$
$\qquad = 3.6 \times 10^6\,joule$

$1HP = 550\,ft.\,lb/sec$ (by definition)
$\qquad = 32.2 \times 550 \times 4.214 \times 10^5\,erg/sec$
$\qquad = 746 \times 10^7\,erg/sec$
$\qquad = 746\,watt$
$1\,MW = 10^6\,watt$
$1\,cal = 1\,calorie = 4.2\,joule$
$1eV = 1.6 \times 10^{-19} joule$
($e = 1.6 \times 10^{-19}$ magnitude of charge on the electron in coulombs)

EXAMPLE 50. An engine of mass 20 tons pulls a train of 20 wagons, each of mass 20 tons with constant velocity of $72\,km/hr$ on a level track. If the coefficient of kinetic friction is $\mu_k = 0.01$, find the power developed by the engine.
APPROACH Use the relation, $P = \vec{F}.\vec{v}$, with $F = \mu_k N$, $N = mg$.
SOLUTION The total forward force F exerted by the engine equals the frictional force $(= \mu_k mg)$ on the whole train (including the engine), as the train is moving forward with constant velocity
The net power developed by the engine
$P = F.v = \mu_k(mg)v$
$\quad = 0.01 \times (420 \times 10^3) \times 10 \times \frac{72 \times 10^3}{3600} = 420\,kW$

EXAMPLE 51. A person decides to use his bath tub water to generate electric power to run a $40\,W$ bulb. The bath tub is located at a height of $10\,m$ from the ground and it holds 200 litre of water. He installs a water-driven wheel generator on the ground. At what rate should the water drain from the bath tub to light the bulb? How long can he keep the bulb on, if the bath tub was full initially? The efficiency of the generator is 90%. ($g = 9.8\,m/s^2$)
APPROACH First of all, calculate the input power to the generator in terms of volume and density of water, then by using efficiency formula write the expression for output power. Finally solve for time t.
SOLUTION If m kg water falls from height h,
$\qquad W = mgh = V\rho gh \qquad [\because m = V\rho]$
Here, V and ρ are volume and density of water.
So, the rate of doing work, i.e., input power
$$P_{in} = \frac{dW}{dt} = \frac{d}{dt}(V\rho gh) = \rho gh \frac{dV}{dt}$$
Now, as efficiency $\eta = \frac{P_{out}}{P_{in}}$, i.e., $P_{out} = \eta P_{in}$
$\Rightarrow 40 = \frac{90}{100} \times \rho gh \frac{dV}{dt}$
or $\quad \frac{dV}{dt} = \frac{40 \times 10}{9 \times 10^3 \times 9.8 \times 10} m^3/sec$
or $\quad \frac{dV}{dt} = \frac{40}{9 \times 9.8} litre/sec = 0.453\,litre/sec$
$[\because 1m^3 = 10^3 litre]$
Further as $V = \left(\frac{dV}{dt}\right) \times t$; so $t = \frac{200}{0.453} = 441$ sec.

EXAMPLE 52. The human heart discharges $75\,cc$ of blood through the arteries at each beat against an average pressure of $10\,cm$ of mercury. Assuming that the pulse frequency is 72 per minute, calculate the rate of working of heart in watt. (density of mercury $= 13.6\,g/cc.$ and $g = 9.8\,m/s^2$).

APPROACH Apply the approach given in the article "Power"

SOLUTION By definition,

Power $(P) = \frac{dW}{dt} = p\frac{dV}{dt}$ $\quad [\because dW = p\, dV]$

Here, $p = h\rho g = 10 \times 13.6 \times 980$
$= 1.3328 \times 10^5\, dynes/cm^2$

and $\frac{dV}{dt}$ = (pulse frequency)
$\times$ (blood discharged per pulse)

i.e., $\frac{dV}{dt} = \frac{72}{60} \times 75 = 90\, cc/sec$

So, power of heart $= 1.3328 \times 10^5 \times 90\, erg/sec$
$\approx 1.19 \times 10^7\, erg/sec = 1.19 W$

EXAMPLE 53. *What is the power output of a $^{238}_{92}U$ reactor if it takes 30 days to use up 2 kg of fuel and if each fission gives 185 MeV of usable energy? (Avogadro number = 6×10^{23} atoms/mol)*

APPROACH *First of all, calculate total number of $^{238}_{92}U$ atoms in 2 kg of fuel by using mole concept. As usable energy released corresponding to each fission is known, therefore calculate total energy released in fission of all $^{238}_{92}U$ atoms. Now, the power output of reactor can be calculated from relation,*

Power output of reactor $= \frac{dW}{dt} = \frac{\text{Total usable energy}}{\text{Time taken}}$

SOLUTION 1 mole, i.e. 235 g of uranium contains 6×10^{23} atoms.
So, $2 kg$, i.e., $2 \times 10^3 g$ of uranium will contain $= \frac{2 \times 10^3 \times 6 \times 10^{23}}{235}$ atoms $= 5.106 \times 10^{24}$ atoms

Now, as in each fission only one uranium atom is consumed, i.e., Energy yield per uranium atom
$= 185\, MeV = 185 \times 10^6\, eV$
$= 185 \times 10^6 \times 1.6 \times 10^{-19} J$
$= 2.96 \times 10^{-1}\, J$

So, energy produced by $2\, kg$ uranium = (No. of atoms)×(energy/atom)
$= 5.106 \times 10^{24} \times 2.96 \times 10^{-11}$
$= 1.514 \times 10^{14} J$

As 2 kg uranium is consumed in 30 days, i.e., $1.514 \times 10^{14} J$ of energy is produced in the reactor in 30 days, i.e., $30 \times 24 \times 60 \times 60\, sec = 2.592 \times 10^6$ sec.

So, Power output of reactor
$= \frac{dW}{dt} = \frac{1.514 \times 10^{14} J}{2.592 \times 10^6 s} = 58.4\, MW$

EXAMPLE 54. *An automobile of mass m accelerates, starting from rest, while the engine supplies constant power P; show that:*
(a) The velocity is given as a function of time by
$v = (2P/m)^{1/2}$
(b) The position is given as a function of time by
$s = (8P/9m)^{1/2} t^{3/2}$

SOLUTION (a) Given that power $P = Fv$ = constant

i.e., $m\frac{dv}{dt}v = P$ $\quad \left[\because F = ma = m\frac{dv}{dt}\right]$

or $\int v\, dv = \int \frac{P}{m} dt$

which on integration gives: $\frac{v^2}{2} = \frac{P}{m} t + C_1$

Now, as initially the body is at rest, i.e., $v = 0$ at $t = 0$, so $0 = 0 + C_1 \Rightarrow C_1 = 0$

$\therefore \quad \frac{v^2}{2} = \frac{P}{m} t$

or $\quad v = \left(\frac{2P}{m}\right)^{1/2}$ ... (1)

(b) By definition, $v = \frac{ds}{dt}$, which in the light of equation (1) becomes

$\frac{ds}{dt} = \left(\frac{2P}{m}\right)^{1/2}$ i.e., $ds = \left(\frac{2Pt}{m}\right)^{1/2} dt$

i.e., $\int ds = \int \left(\frac{2Pt}{m}\right)^{1/2} dt$

which on integration, gives

$s = \left(\frac{2P}{m}\right)^{1/2} \frac{2}{3} t^{3/2} + C_2$

Now, as at $t = 0$, $s = 0$ so $0 = 0 + C_2 \Rightarrow C_2 = 0$

$\therefore s = \left(\frac{8P}{9m}\right)^{1/2} t^{3/2}$

EXAMPLE 55. *Find the maximum energy stored in the spring shown in the figure, for which the block remains stationary on the rough horizontal surface. The coefficient of static friction between the block and horizontal surface is μ.*

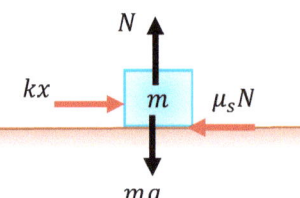

APPROACH Since, elastic potential energy stored in spring block system is given by $U = \frac{1}{2}kx^2$, therefore, corresponding to maximum value of U, the compression in the spring must be maximum.

At maximum compression, the spring force on the block will also be maximum and the block must be in limiting equilibrium under the spring force and the maximum possible force of static friction.

i.e., $kx = f_{max} = \mu N$, with $N = mg$

From this relation, find the compression in spring, x and then use it in $U = \frac{1}{2}kx^2$ to get the maximum potential energy stored in the spring.

SOLUTION Let the spring be compressed by distance x. The potential energy stored in the spring

$U = \frac{1}{2}kx^2$.

From FBD, in equilibrium, we have
In vertical direction:
$N - Mg = 0 \Rightarrow N = mg$ ... (1)
In horizontal direction: $kx - f_{max} = 0$
$\Rightarrow kx = f_{max} = \mu N$ ... (2)
Using the value of N from (1) in (2), we get
Therefore, $f_{max} = \mu mg$

$$kx = \mu mg$$
$$\Rightarrow x = \frac{\mu mg}{k}$$

The maximum potential energy stored in the spring
$$U_{max} = \frac{1}{2}kx^2 = \frac{1}{2}k\left(\frac{\mu mg}{k}\right)^2 = \frac{\mu^2 m^2 g^2}{2k}$$

EXAMPLE 56. A variable force $\vec{F}$ is maintained tangent to a frictionless, semicircular surface (see adjoining figure). By slow variations in the force, a block with weight w is moved, and the spring to which it is attached is stretched from position 1 to position 2. The spring has negligible mass and force constant k. The end of the spring moves in an arc of radius a. Calculate the work done by the force $\vec{F}$.

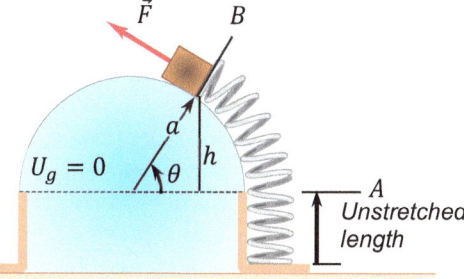

APPROACH We use the fact that, the work done by a non conservative force = change in total mechanical energy of the system. Also note that variation of force F is very slow, it means that the block always remains in equilibrium during the motion. Therefore, the change in kinetic energy will be zero.

SOLUTION Total linear expansion in the spring, $s = a\theta$
Now, draw a perpendicular from position B on the horizontal line passing through the position A. If the height of this perpendicular is h, then from adjoining figure, we have-
$$\sin\theta = \frac{h}{a} \qquad \dots (1)$$
or $\quad h = a\sin\theta$

The net work done by force F = change in mechanical energy
= Change in PE + Change in KE
= (change in gravitational PE + change in elastic PE) + Change in KE
$= (mgh - 0) + \frac{1}{2}ks^2 - 0$

[We have considered the reference gravitational energy level at the horizontal line passing through A]
$= mga\sin\theta + \frac{1}{2}ka^2\theta^2$

30. CHECKPOINT 4

1. ●●In adjoining figure, a chain is held on a frictionless table with one fourth of its length hanging over the edge. If the chain has length $L = 28$ cm and mass $m = 0.012$ kg, how much work is required to pull the hanging part back onto the table?

2. ●●A 1.20 kg block slides down a frictionless incline with a slope angle of 45°, starting from a height $h = 2.30\ m$ above the bottom of the incline, as shown in following figure. The incline meets a frictionless horizontal surface, at the end of which is a spring in its equilibrium position ($k = 460.\ N/m$) used to stop the block. Find the maximum compression of the spring.

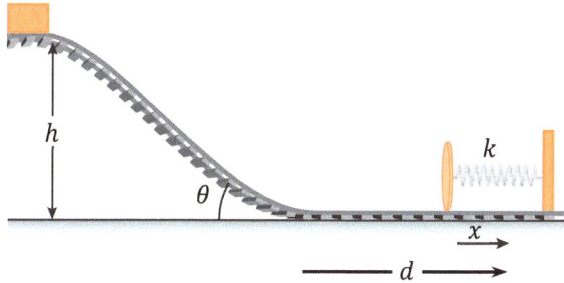

3. ●●●A chain of length l and mass m lies on the surface of a smooth sphere of radius $R > l$ with one end tied to the top of the sphere.
   (a) Find the gravitational potential energy of the chain with reference level at the centre of the sphere.
   (b) Suppose the chain is released and slides down the sphere. Find the kinetic energy of the chain, when it has slid through an angle α
   (c) Find the tangential acceleration $\frac{dv}{dt}$ of the chain when the chain starts sliding down.

4. ●●●The flexible chain of length $\frac{\pi r}{2}$ and mass per unit length λ is released from rest with $\theta = 0°$ in the smooth circular channel and falls through the hole in the supporting surface. Determine the velocity v of the chain as the last link leaves the slot.

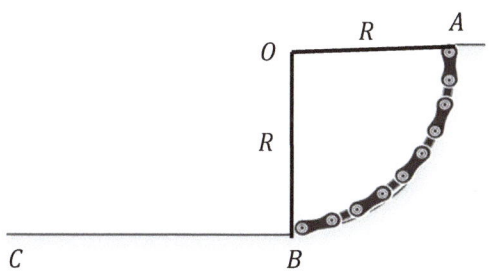

5. •••The flexible chain of length $\frac{\pi r}{2}$ and mass per unit length λ is released from rest with $\theta = 0°$ in the smooth circular channel and falls through the hole in the supporting surface. Determine the velocity v of the chain as the last link leaves the slot.

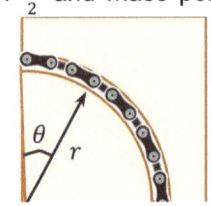

6. •••A chain of length L and mass M is arranged as shown in following five cases. The *correct decreasing* order of potential energy (assumed zero at horizontal surface) is.

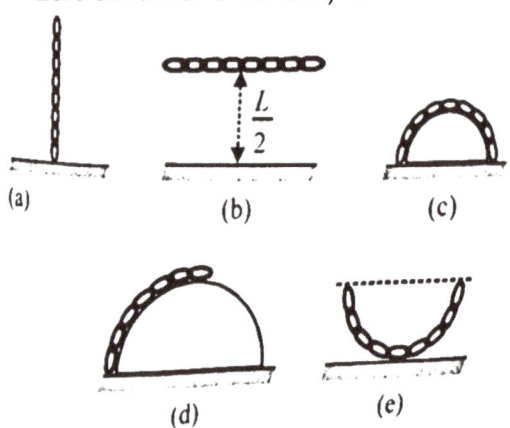

7. ••Discuss the motion of a particle and draw conservative force vs position graph corresponding to hypothetical potential energy – position graph shown in following figure.

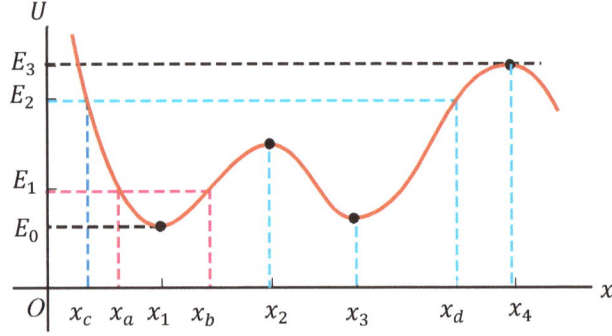

8. •A factory uses a motor and a cable to drag a $300\ kg$ machine to the proper place on the factory floor. What power must the motor supply to drag the machine at a speed of $0.50\ m/s$? The coefficient of friction between the machine and the floor is 0.60.

9. ••A $1500\ kg$ car has a front profile that is $1.6\ m$ wide by $1.4\ m$ high and air drag $655N$. The coefficient of rolling friction is 0.02. What power must the engine provide to drive at a speed $30\ m/s$ if 25% of the power is "lost" before reaching the drive wheels?

31. SOLVED EXAMPLES

1. Find the work done on a particle of mass m by a force, $K\left[\frac{x}{(x^2+y^2)^{3/2}}\hat{i} + \frac{y}{(x^2+y^2)^{3/2}}\hat{j}\right]$ (K being a constant of appropriate dimensions), when the particle is taken from the point $(a, 0)$ to the point $(0, a)$ along a circular path of radius a about the origin in the x-y plane.

SOLUTION Let $\vec{r} = x\hat{i} + y\hat{j}$, then $r = \sqrt{x^2 + y^2}$

$\therefore \vec{F} = K\left[\frac{x\hat{i}+y\hat{j}}{\left(\sqrt{x^2+y^2}\right)^3}\right] = K\frac{\vec{r}}{r^3}$

Thus, $\vec{F}$ is pointed along the radial vector away from the origin. As, the radial vector is always perpendicular to the given circular path, therefore, corresponding to each small displacement $d\vec{s}$ along the circular path, work done is $dW = \vec{F}.d\vec{s} = 0$. Hence, the total work done by the force $\vec{F}$ is zero.

2. A block of mass $2\ kg$ is free to move along the x axis. It is at rest and from $t = 0$ onwards it is subjected to a time dependent force $F(t)$ in the x direction. The force $F(t)$ varies with t as shown in the figure. Calculate the kinetic energy of the block after 4.5 seconds is

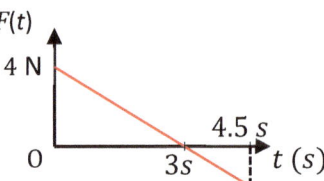

SOLUTION The t-F diagram is a straight line passing through
(0,4) and (3,0). The equation of this straight line is
$F = -\frac{4}{3}t + 4$
Newton's second law gives acceleration of the block as
$$a = \frac{F}{m} = -\frac{2}{3}t + 2$$
Integrate to get the velocity v at 4.5 s
$$v = \int_0^{4.5} a dt = \int_0^{4.5}\left(-\frac{2}{3}t + 2\right)dt = 2.25 m/s$$
Thus, kinetic energy of the block at 4.5 s is
$$K = \frac{1}{2}mv^2 = 5.06 J$$

3. A block (B) is attached to two unstretched spring S_1 and S_2 with spring constants k and $4k$, respectively (see figure *I*). The other ends are attached to identical supports M_1 and M_2 not attached to the walls. The springs and supports have negligible masses. There is no friction anywhere. The block

B is displaced towards wall 1 by a small distance x (figure II) and released. The block returns and moves a maximum distance y towards wall 2. Displacements x and y are measured with respect to the equilibrium position of the block B. What is the ratio $\frac{y}{x}$?

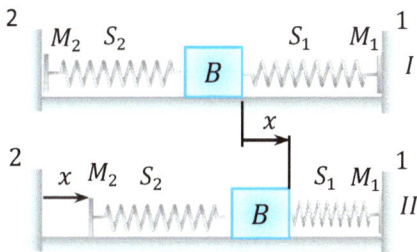

SOLUTION When B is displaced towards the wall 1 by a distance x, S_1 is compressed by x and S_2 is unstretched. The total energy of the spring-mass system is
$$E_1 = \frac{1}{2}kx^2$$
When B returns and moves a maximum distance y, S_2 is compressed by y and S_1 is unstretched. The total energy of the system is
$$E_2 = \frac{1}{2}(4k)y^2$$
Since the springs and the supports have negligible masses and there is no friction, total energy of the system is conserved i.e.,
$$\frac{1}{2}kx^2 = \frac{1}{2}(4k)y^2$$
which gives, $\frac{y}{x} = \frac{1}{2}$.

4. A wind powered generator converts wind energy into electrical energy. Assume that the generator converts a fixed fraction of the wind energy intercepted by its blades into electrical energy. For wind speed v, find the electrical power output produced by the generator.

SOLUTION. If the blades of the generator sweep a circular area A in space. then the intercepted air can be modelled as a cylinder of cross-sectional area A moving with a constant velocity v as shown in following Fig.1. The volume of air intercepted by the blades of generator in time Δt, is given by-

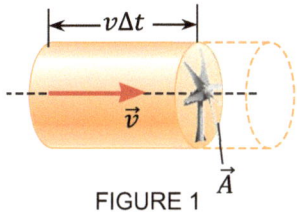

FIGURE 1

$$\Delta V = Av\Delta t$$
This volume is shown by shaded part in Fig. 1.

If the density of air is ρ, then the mass of this cylindrical air volume,
$$\Delta m = \rho \Delta V = \rho A v \Delta t$$
Therefore, the kinetic energy of this cylindrical air section before striking the blades of the generator,
$$K = \frac{1}{2}(\Delta m)v^2 = \frac{1}{2}(\rho A v \Delta t)v^2$$
Suppose, after striking the blades of generator, the air comes to rest, then change in Kinetic energy of the air disk
$$K_f - K_i = 0 - \frac{1}{2}(\rho A v \Delta t)v^2 = -\frac{1}{2}(\rho A \Delta t)v^3$$
Negative sign show that there is a loss of kinetic energy of the air which is transferred to the generator in the form of rotational kinetic energy of the blades. Therefore, the rotational kinetic energy gained by the blades of generator
$$\Delta K = \frac{1}{2}(\rho A \Delta t)v^3$$
Work done by air, in time Δt, corresponding to this transfer of kinetic energy
$$\Delta W = \Delta K = \frac{1}{2}(\rho A \Delta t)v^3$$
∴ Power input to generator corresponding to above work
$$P = \frac{\Delta W}{\Delta t} = \frac{1}{2}\rho A v^3$$
If the generator is 100% efficient, then the electric power output of generator $= \frac{1}{2}\rho A v^3$

5. Consider an elliptically shaped rail PQ in the vertical plane with $OP = 3\,m$ and $OQ = 4\,m$. A block of mass $1\,kg$ is pulled along the rail from P to Q with a force of $18\,N$, which is always parallel to line PQ (see figure). Assuming no frictional losses, find the kinetic energy of the block when it reaches to Q [Given $g = 10\,m/s^2$.]

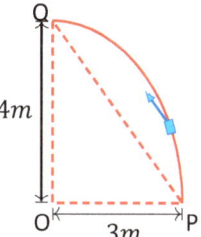

SOLUTION Consider a small displacement $d\vec{s}$ along the elliptical path. Resolve $d\vec{s}$ in directions parallel and perpendicular to $\vec{F}$, say $d\vec{s}_{\parallel}$ and $d\vec{s}_{\perp}$. The work done by $\vec{F}$ for the path from P to Q is
$$W = \int \vec{F}.d\vec{s} = \int \vec{F}.d\vec{s}_{\parallel}$$
$$= F \int ds_{\parallel} = 18(5) = 90\,J \quad \ldots (1)$$
By work-energy theorem
$$W = \Delta K + \Delta U \quad \ldots (2)$$
where $\Delta U = U_f - U_i$ is the change in the potential energy and ΔK is the change in the kinetic energy. From equations (1) and (2),
$$\Delta K = W - mgh = 90 - 1(10)4 = 50\,J.$$

6. A particle of mass $0.2\,kg$ is moving in one dimension under a force that delivers a constant

power $0.5\ W$ to the particle. If the initial speed of the particle is zero, find the speed (in m/s) after $5\ s$.

SOLUTION The power P delivered by a force $\vec{F}$ is related to velocity $\vec{v}$ by
$$P = \vec{F}\cdot\vec{v}. \qquad \ldots (1)$$
From Newton's second law,
$$\vec{F} = m\frac{d\vec{v}}{dt}. \qquad \ldots (2)$$
In one-dimension, equations (1) and (2) give
$$P = mv\frac{dv}{dt} \qquad \ldots (3)$$
Integrate equation (3) with time (P is constant) to get
$$v^2 = v_0^2 + \frac{2Pt}{m}.$$
Substitute the values of v_0, P, t and m to get $v = 5\ m/s$.

7. A freight company uses a compressed spring to shoot $2.0\ kg$ packages up a $1.0\ m$-high frictionless ramp into a truck, as shown in following figure. The spring constant is $500\ N/m$ and the spring is compressed $30\ cm$.

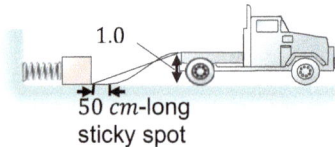

(a) What is the speed of the package when it reaches the truck?

(b) A careless worker spills his soda on the ramp. This creates a 50-cm-long sticky spot with a coefficient of kinetic friction 0.30. Will the next package make it into the truck?

SOLUTION We will use the spring, the package, and the ramp as the system. We will model the package as a particle
We place the origin of our coordinate system on the end of the spring when it is compressed and is in contact with the package to be shot
(a) The energy conservation equation is
$$K_1 + U_{g1} + U_{s1} + \Delta E_{th} = K_0 + U_{g0} + U_{s0} + W_{ext}$$
$$\frac{1}{2}mv_1^2 + mgy_1 + \frac{1}{2}k(x_e - x_e)^2 + \Delta E_{th}$$
$$= \frac{1}{2}mv_0^2 + mgy_0 + \frac{1}{2}k(\Delta x)^2 + W_{ext} = 0J$$
(the frictionless ramp), $b = 0m/s$, $V_0 = 0m$, $\Delta x = 30$cm, and $W_{ext} = 0J$, we get
$$\frac{1}{2}mr_1^2 + mg(1.0m) + 0J + 0J = 0J + 0J + \frac{1}{2}k(0.30m)^2$$
$$\frac{1}{2}(2.0\text{kg})V_1^2 + (2.0\text{kg})\left(\frac{9.8\text{m}}{s^2}\right)(1.0m)$$
$$= \frac{1}{2}(500\text{N/m})(0.30m)^2$$
$$V_1 = 1.7\text{m/s}$$
(b) How high can the package go after crossing the sticky spot? If the package can reach $y_1 \geq 1.0\ m$ before stopping, $(v_1 = 0)$ then it makes it. But if $y_1 < 1.0\ m$ when $v_1 = 0$, the package does not make it. The friction of the sticky spot generates thermal energy
$$\Delta E_{th} = (\mu_k mg)\Delta x = (0.30)(2.0\text{kg})\left(\frac{9.8\text{m}}{s^2}\right)(0.50m)$$
$$= 2.94J$$
The energy conservation equation is now
$$\frac{1}{2}mv_1^2 + mgy_1 + \Delta E_{th} = \frac{1}{2}k(\Delta x)^2$$
If we set $v_1 = 0\ m/s$ to find the highest point the package can reach, we get
$$y_1 = \left(\frac{1}{2}k\Delta x^2 - \Delta E_{th}\right)/mg$$
$$= \left[\frac{1}{2}\left(\frac{500\text{N}}{m}\right)(0.30m)^2 - 2.94J\right]/[(2.0\text{kg})(9.8\ m/s^2)]$$
$$= 0.998\ m$$

8. A 6000-kg freight car rolls along rails with negligible friction. The car is brought to rest by a combination of two coiled springs as illustrated in following figure. Both springs are described by Hooke's law and have spring constants $k_1 = 1600\ N/m$ and $k_2 = 3400\ N/m$. After the first spring compresses a distance of $30.0\ cm$, the second spring acts with the first to increase the force as additional compression occurs as shown in the graph. The car comes to rest $50.0\ cm$ after first contacting the two-spring system. Find the car's initial speed.

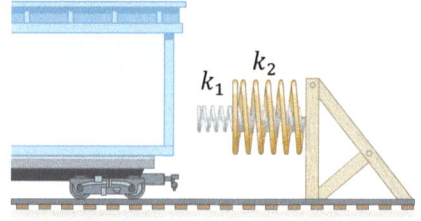

(a)

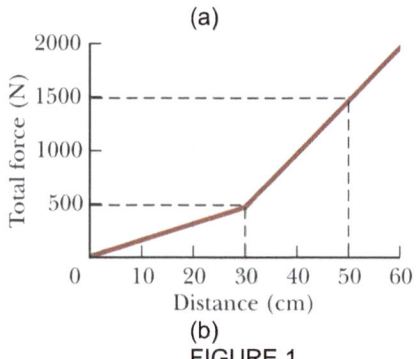

(b)

FIGURE 1

SOLUTION Compare an initial picture of the rolling car with a final picture with both springs compressed. From conservation of energy, we have
$$K_1 + \Sigma W = K_f$$
Work by both springs changes the car's kinetic energy
$$K_1 + \frac{1}{2}k_1(x_{1i}^2 - x_{1f}^2) + \frac{1}{2}k_2(x_{2i}^2 - x_{2f}^2) = K_f$$
Substituting,
$$\frac{1}{2}mv_i^2 + 0 - \frac{1}{2}(1600N/m)(0.500m)^2$$

$$+0 - \frac{1}{2}(3400\text{N/m})(0.200\text{m})^2 = 0$$

Which gives
$$\frac{1}{2}(6000\text{kg})v_i^2 - 200\text{J} - 68.0\text{J} = 0$$

Solving for v_i,
$$v_1 = \sqrt{\frac{2(268\text{J})}{6000\text{kg}}} = 0.299 \text{m/s}$$

A particle of mass $m = 1.18\ kg$ is attached between two identical springs on a frictionless, horizontal tabletop. Both springs have spring constant k and are initially unstressed, and the particle is at $x = 0$. (a) The particle is pulled a distance x along a direction perpendicular to the initial configuration of the springs as shown in following figure. Show that the force exerted by the springs on the particle is

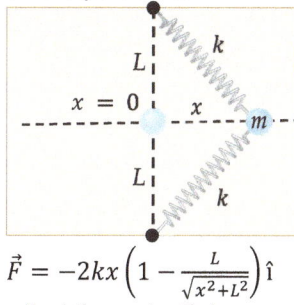

$$\vec{F} = -2kx\left(1 - \frac{L}{\sqrt{x^2+L^2}}\right)\hat{\imath}$$

(b) Show that the potential energy of the system is
$$U(x) = kx^2 + 2kL(L - \sqrt{x^2 + L^2})$$

SOLUTION (a) The new length of each spring is $\sqrt{x^2 + L^2}$, so its extension is $\sqrt{x^2 + L^2} - L$ and the force it exerts is $k(\sqrt{x^2 + L^2} - L)$ toward its fixed end. The y components of the two spring forces add to zero. Their x components (with $\cos\theta = \frac{x}{\sqrt{x^2+L^2}}$) add to
$$\vec{F} = -2k(\sqrt{x^2 + L^2} - L)\frac{x}{\sqrt{x^2+L^2}}\hat{\imath}$$
$$= -2kx\left(1 - \frac{L}{\sqrt{x^2+L^2}}\right)\hat{\imath}$$

(b) Choose $U = 0$ at $x = 0$. Then at any point the potential energy of the system is
$$U(x) = -\int_0^x F_x dx = -\int_0^x \left(-2kx + \frac{2kLx}{\sqrt{x^2+L^2}}\right)dx$$
$$= 2k\int_0^x x\,dx - 2kL\int_0^L \frac{x}{\sqrt{x^2+L^2}}dx$$
$$U(x) = kx^2 + 2kL(L - \sqrt{x^2 + L^2})$$

9. At a construction site, a 65.0-kg bucket of concrete hangs from a light (but strong) cable that passes over a light, friction-free pulley and is connected to an 80.0-kg box on a horizontal roof (see adjoining figure). The cable pulls horizontally on the box, and a 50.0-kg bag of gravel rests on top of the box. The coefficients of friction between the box and roof are shown. (a) Find the friction force on the bag of gravel and on the box. (b) Suddenly a worker picks up the bag of gravel. Use energy conservation to find the speed of the bucket after it has descended 2.00 m from rest.

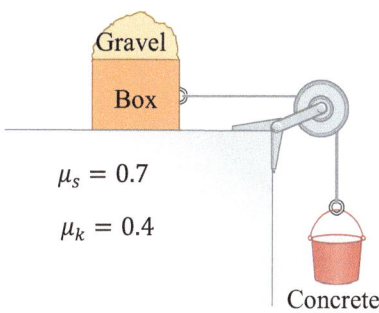

APPROACH Apply $\sum \vec{F} = m\vec{a}$ to the bag and to the box. Apply Eq. $K_1 + U_{grav,1} + W_{othe} = K_2 + U_{grav,2}$ to the motion of the system of the box and bucket after the bag is removed.

SOLUTION Let $y = 0$ the final height of the bucket, so $y_1 = 2.00\ m$ and $y_2 = 0$. $K_1 = 0$. The box and the bucket move with the same speed v, so $K_2 = \frac{1}{2}(m_{box} + m_{bucket})v^2$. $W_{ot} = -f_k d$, with $d = 2.00m$ and $f_k = \mu_k m_{box} g$. Before the bag is removed, the maximum possible friction force the roof can exert on the box is $(0.7)(80.0\ kg + 50.0\ kg)(9.80\ m/s^2) = 892\ N$. This is larger than the weight of the bucket (637 N), so before the bag is removed the system is at rest.

(a) The friction force on the bag of gravel is zero, since there is no other horizontal force on the bag for friction to oppose. The static friction force on the box equals the weight of the bucket, 637 N

(b) $m_{bucket} g y_1 - f_k d = \frac{1}{2} m_{tot} v^2$, with $m_{tot} = 145.0\ kg$.

$$\therefore v = \sqrt{\frac{2}{m_{tot}}(m_{bucket} g y_1 - \mu_k m_{box} g d)}$$

$$v = \sqrt{\frac{2}{145.0\ kg}[(65.0\ kg)(9.8\ m/s^2)(2.00m) - (0.4)(80.0\ kg)(9.8\ m/s^2)(2.00\ m)]}$$

$$v = 2.99\ m/s$$

☞ If we apply $\sum \vec{F} = m\vec{a}$ to the box and to the bucket we can calculate their common acceleration a. Then a constant acceleration equation applied to either object gives $v = 2.99\ m/s$, in agreement with our result obtained using energy methods.

10. A 3.00-kg fish is attached to the lower end of a vertical spring that has negligible mass and force constant. The spring initially is neither stretched nor compressed. The fish is released from rest. (a) What is its speed after it has descended 0.05 m from its initial position? (b) What is the maximum speed of the fish as it descends?

APPROACH: Only conservative forces (gravity and the spring force) act on the fish, so its mechanical energy is conserved.
According to energy conservation tells, we have
$K_1 + U_1 + W_{other} = K_2 + U_2$, where $W_{other} = 0$. $U_g = mgy$, $K = \frac{1}{2}mv^2$, and $U_{spring} = \frac{1}{2}ky^2$

SOLUTION (a) $K_1 + U_1 + W_{other} = K_2 + U_2$
Let y be the distance the fish has descended, so $y = 0.05\,m$.
$K_1 = 0, W_{other} = 0, U_1 = mgy, K_2 = \frac{1}{2}mv_2^2$, and $U_2 = \frac{1}{2}ky^2$. Solving for K_2 gives
$K_2 = U_1 - U_2 = mgy - \frac{1}{2}ky^2$
$= (3.00kg)(9.8m/s^2)(0.0500m)$
$\quad - \frac{1}{2}(900N/m)(0.0500m)^2$
$K_2 = 1.47J - 1.125J = 0.345J$
Solving for v_2 gives
$v_2 = \sqrt{\frac{2K_2}{m}} = \sqrt{\frac{2(0.345J)}{3.00kg}} = 0.48 m/s$

(b) The maximum speed is when K_2 is maximum, which is when $\frac{dK_2}{dy} = 0$. Using $K_2 = mgy - \frac{1}{2}ky^2$ gives
$\frac{dK_2}{dy} = mg - ky = 0$
Solving for y gives $y = \frac{mg}{k} = \frac{(3.00\,kg)(9.8m/s^2)}{900N/m} = 0.03267\,m$. At this y,
$K_2 = (3.00kg)\left(\frac{9.8m}{s^2}\right)(0.03267m)$
$\quad - \frac{1}{2}(900N/m)(0.03267m)^2$
$K_2 = 0.9604J - 0.4803J = 0.4801J$, so
$v_2 = \sqrt{\frac{2K_2}{m}} = 0.566 m/s$

☞ The speed in part (b) is greater than the speed in part (a), as it should be since it is the maximum speed.

11. A certain spring is found *not* to obey Hooke's law; it exerts a restoring force $F_x(x) = -\alpha x - \beta x^2$ if it is stretched or compressed, where $\alpha = 60.0\,N/m$ and $\beta = 18.0\,N/m^2$. The mass of the spring is negligible. (a) Calculate the potential-energy function $U(x)$ for this spring. Let $U = 0$ when $x = 0$. (b) An object with mass $0.90\,kg$ on a frictionless, horizontal surface is attached to this spring, pulled a distance $1.00\,m$ to the right (the $+ve$ direction) to stretch the spring, and released. What is the speed of the object when it is $0.50\,m$ to the right of the $x = 0$ equilibrium position?

(a) **APPROACH** Use Eq. $W = \int F_x(x)dx$ to calculate W and then use $W = -\Delta U$ to identify the potential energy function $U(x)$

$W_{F_x} = U_1 - U_2 = \int_{x_1}^{x_2} F_x(x)dx$
Let $x_1 = 0$ and $U_1 = 0$. Let x_2 be some arbitrary point x, so $U_2 = U(x)$.

SOLUTION $U(x) = -\int_0^x F_x(x)dx$
$= -\int_0^x (-\alpha x - \beta x^2)\,dx = \int_0^x (\alpha x + \beta x^2)\,dx$
$= \frac{1}{2}\alpha x^2 + \frac{1}{3}\beta x^3$

☞ If $\beta = 0$, the spring does obey Hooke's law, with, $k = \alpha$ and our result reduces to $\frac{1}{2}kx^2$.

(b) **APPROACH** Apply conservation of energy to the motion of the object.
SOLUTION The system at points 1 and 2 is sketched in following figure
$K_1 + U_1 + W_{other} = K_2 + U_2$

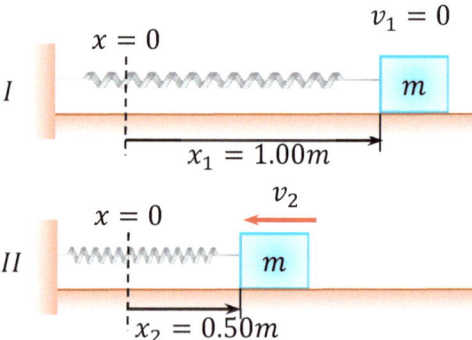

The only force that does work on the object is the spring force, so $W_{othe} = 0$.
CALCULATION: $K_1 = 0, K_2 = \frac{1}{2}mv_2^2$
$U_1 = U(x_1) = \frac{1}{2}\alpha x_1^2 + \frac{1}{3}\beta x_1^3$
$= \frac{1}{2}(60.0N/m)(1.00m)^2 + \frac{1}{3}(18.0N/m^2)(1.00m)^3$
$= 36.0J$

$U_2 = U(x_2) = \frac{1}{2}\alpha x_2^2 + \frac{1}{3}\beta x_2^3$
$= \frac{1}{2}(60.0N/m)(0.500m)^2 + \frac{1}{3}(18.0N/m^2)(0.500m)^3$
$= 8.25J$
Thus, $36.0J = \frac{1}{2}mv_2^2 + 8.25J$
$v_2 = \sqrt{\frac{2(36.0J - 8.25J)}{0.900kg}} = 7.85 m/s$

☞ The elastic potential energy stored in the spring decreases and the kinetic energy of the object increases.

12. A pump motor is used to deliver water at a certain rule from a given pipe. To obtain 'n' times water from the same pipe in the same time by what amount (a) the force and (b) power of the motor should be increased?

SOLUTION If a liquid of density ρ is flowing through a pipe of cross-section A at speed v the mass coming out per sec will be

$\frac{dm}{dt} = Av\rho$

So, to get n times water in the same time

$\left(\frac{dm}{dt}\right)' = n\left(\frac{dm}{dt}\right)$

i.e., $A'v'\rho' = nAv\rho$

but, as pipe and liquid are same, $\rho' = \rho$ and $A' = A$

$\therefore \quad v' = nv$ ... (1)

(a) Now as, $F = v\frac{dm}{dt}$

$\frac{F'}{F} = \frac{v'(dm/dt)'}{v(dm/dt)} = \frac{nv(n\,dm/dt)}{v(dm/dt)} = n^2$

or $\quad F' = n^2 F$ ... (2)

(b) And as $P = Fv$

$\frac{P'}{P} = \frac{F'v'}{Fv} = \frac{(n^2F)(nv)}{Fv} = n^3$

i.e., $\quad P' = n^3 P$

So, to get n times water, force must be increased n^2 times while power n^3 times.

32. QUESTIONS AND EXERCISES

32.1. CONCEPTUAL QUESTIONS

1. Is it possible to do work on an object that remains at rest?
2. A friend makes the statement, "Only the total force acting on an object can do work." Is this statement true or false? If it is true, state why; if it is false, give a counterexample.
3. A friend makes the statement, "A force that is always perpendicular to the velocity of a particle does no work on the particle." Is this statement true or false? If it is true, state why; if it is false, give a counterexample.
4. A process occurs in which a system's potential energy decreases while the system does work on the environment. Does the system's kinetic energy increase, decrease, or stay the same? Or is there not enough information to tell? Explain
5. A process occurs in which a system's potential energy increases while the environment does work on the system. Does the system's kinetic energy increase, decrease, or stay the same? Or is there not enough information to tell? Explain.
6. The kinetic energy of a system decreases while its potential energy and thermal energy are unchanged. Does the environment do work on the system, or does the system do work on the environment? Explain.
7. You drop a ball from a high balcony and it falls freely. Does the ball's kinetic energy increase by equal amounts in equal time intervals, or by equal amounts in equal distances? Explain
8. A woman swimming upstream is not moving with respect to the shore. Is she doing any work? If she stops swimming and merely floats, is work done on her?
9. A particle moves in a vertical plane along the *closed* path seen in following figure, starting at A and eventually returning to its starting point. How much work is done on the particle by gravity? Explain.

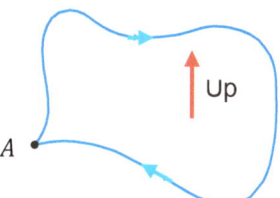

10. Can a centripetal force ever do work on an object? Explain.
11. Why is it tiring to push hard against a solid wall even though you are doing no work?
12. Does the work done by a force depend on the choice of coordinate system?
13. Can the normal force on an object ever do work? Explain.
14. Two bullets are fired at the same time with the same kinetic energy. If one bullet has twice the mass of the other, which has the greater speed and by what factor? Which can do the most work?
15. Does the net work done on a particle depend on the choice of reference frame? How does this affect the work-energy principle?
16. You lift a heavy book from a table to a high shelf. List the forces on the book during this process, and state whether each is conservative or nonconservative.
17. The net force acting on a particle is conservative and increases the kinetic energy by $300\,J$. What is the change in (a) the potential energy, and (b) the total energy, of the particle?
18. Is it possible for the kinetic energy of an object to be negative? Is it possible for the gravitational potential energy of an object to be negative? Explain.
19. When a "superball" is dropped, can it rebound to a greater height than its original height?

20. Can the work by kinetic/static friction on an object be positive? negative? a Zero?
21. Why is it tiring to push hard against a solid wall even though no work is done?
22. Does the work done by the net force acting on a particle depend on the (inertial) reference frame of the observer? Does the change in kinetic energy so depend? If so, give examples.
23. A coil spring of mass m rests upright on a table. If you compress the spring by pressing down with your hand and then release it, can the spring leave the table? Explain using the law of conservation of energy.
24. Can kinetic energy of a system be increased without applying any external force on the system?
25. What happens to the gravitational potential energy when water at the top of a waterfall falls to the pool below?
26. A compressed spring is clamped in its compressed position and is then dissolved in acid. What becomes of the spring's potential energy?
27. In picking up a book from the floor and putting it on a table, you do work. However, the kinetic energy of the book does not change. Is there a violation of the work–energy theorem here? Explain why or why not.
28. (a) Where does the kinetic energy come from when a car accelerates uniformly starting from rest? (b) How is the increase in kinetic energy related to the friction force the road exerts on the tires?
29. Does the work–energy theorem hold if friction acts on an object? Explain your answer
30. (a) Can a body have kinetic energy without having momentum? (b) Can a body have momentum without having kinetic energy?
31. The Earth is closest to the Sun in winter (Northern Hemisphere). When is the gravitational potential energy the greatest?
32. Two disks are connected by a stiff spring. Can you press the upper disk down enough so that when it is released it will spring back and raise the lower disk off the table (see adjoining figure)? Can mechanical energy be conserved in such a case?

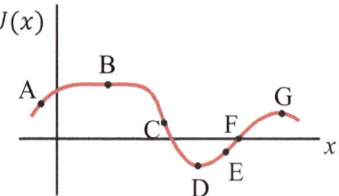

33. Can the total mechanical energy ever be negative? Explain.
34. Suppose you lift a suitcase from the floor to a table. The work you do on the suitcase depends on which of the following: (a) whether you lift it straight up or along a more complicated path, (b) the time the lifting takes, (c) the height of the table, and (d) the weight of the suitcase?
35. Engine 1 produces twice the power of engine 2. Is it correct to conclude that engine 1 does twice as much work as engine 2? Explain.
36. Following figure shows a potential energy curve, (a) At which point does the force have greatest magnitude? (b) For each labeled point, state whether the force acts to the left or to the right, or is zero. (c) Where is there equilibrium and of what type is

37. (a) Describe in detail the velocity changes of a particle that has energy E_3 in following figure as it moves from x_6 to x_5 and back to x_6 (b) Where is its kinetic energy the greatest and the least?

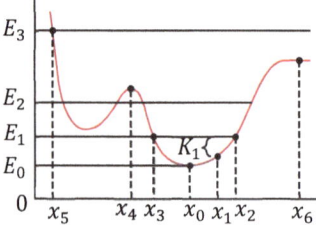

38. Name the type of equilibrium for each position of the balls in the following figure.

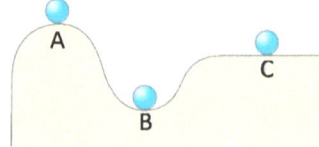

39. You often hear that the energy we use on the earth comes from the sun. Explain how this is true. For example, how does the kinetic energy of a running person come from the sun?

32.2. PROBLEMS

1. •A 106-kg object is initially moving in a straight line with a speed of 51.3 m/s. (a) If it is brought to a stop with a deceleration of 1.97 m/s², what force is required, what distance does the object travel, and how much work is done by the force? (b) Answer the same questions if the object's deceleration is 4.82 m/s².

2. •How much work is done by the gravitational force when a 280-kg pile driver falls 2.80 m?

3. •A 47.2-kg block of ice slides down an incline 1.62 m long and 0.902 m high. A worker pushes up on the ice parallel to the incline so that it slides down at constant speed. The coefficient of kinetic friction between the ice and the incline is 0.11. Find (a) the force exerted by the worker, (b) the work done by the worker on the block of ice, and (c) the work done by gravity on the ice. (Take $g = 9.81 \, m/s^2$)

4. •Electric fields can be used to pull electrons out of metals. To remove an electron from tungsten, the electric field must do 4.5 eV of work. Suppose that the distance over which the electric field acts is 3.4 nm. Calculate the minimum force that the field must exert on the electron being removed.

5. •How high will a 1.85-kg rock go if thrown straight up by someone who does 80.0 J of work on it? Neglect air resistance.

6. •A cord is used to lower vertically a block of mass M a distance d at a constant downward acceleration of g/4. (a) Find the work done by the cord on the block. (b) Find the work done by the force of gravity.

7. •A worker can lift a 75-kg block directly off the ground on to a loading dock, or can push the block up a frictionless incline. from the ground to the loading dock. Lifting the block requires 680 J of work to be done. Pushing the block up the incline requires a minimum applied force of 320 N. Find the angle the incline makes with the horizontal.

8. ••A 75.0-kg painter climbs a ladder that is 2.75 m long leaning against a vertical wall. The ladder makes a 30° angle with the wall. (a) How much work does gravity do on the painter? (b) Does the answer to part (a) depend on whether the painter climbs at constant speed or accelerates up the ladder?

9. •A 75.0-kg firefighter climbs a flight of stairs 20.0 m high. How much work is required?

10. •Eight books, each 4.0 cm thick with mass 1.8 kg, lie flat on a table. How much work is required to stack them one on top of another?

11. ••You apply a constant force $\vec{F} = (-68.0 \, N)\hat{\imath} + (36.0 \, N)\hat{\jmath}$ to a 380-kg car as the car travels 48.0 m in a direction that is 240° counter-clockwise from the x-axis. How much work does the force you apply do on the car?

12. •In pedalling a bicycle uphill, a cyclist exerts a downward force of 450 N during each stroke. If the diameter of the circle traced by each pedal is 36 cm, calculate how much work is done in each stroke.

13. ••A box is sliding with a speed of 4.50 m/s on a horizontal surface when, at point P, it encounters a rough section. On the rough section, the coefficient of friction is not constant, but starts at 0.1 at P and increases linearly with distance past P, reaching a value of 0.6 at 12.5 m past point P. (a) Use the work–energy theorem to find how far this box slides before stopping. (b) What is the coefficient of friction at the stopping point? (c) How far would the box have slid if the friction coefficient didn't increase but instead had the constant value of 0.1?

14. •A spring has $k = 65 \, N/m$. Draw a graph like that in adjoining figure and use it to determine the work needed to stretch the spring from x = 3.0 cm to x = 6.5 cm, where x = 0 refers to the spring's unstretched length.

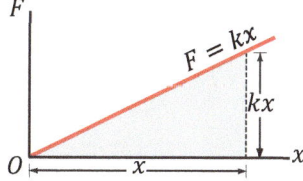

15. ••The net force exerted on a particle acts in the positive x direction. Its magnitude increases linearly from zero at x = 0 to 380 N at x = 3.0m It remains constant at 380 N from x = 3.0 m to x = 7.0 m and then decreases linearly to zero at x = 12.0 m. Determine the work done to move the particle from x = 0 to x = 12 m graphically, by determining the area under the F_x versus x graph.

Sol. See the graph of force vs. distance. The work done is the area under the graph. It can be found from the formula for a trapezoid

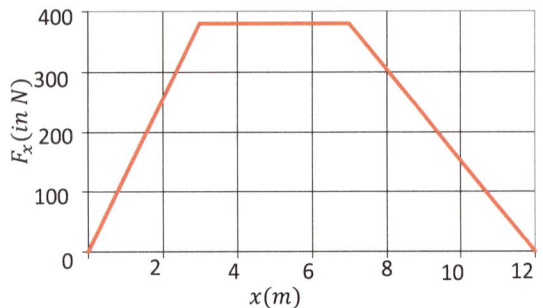

$$W = \frac{1}{2}(12.0\,m + 4.0\,m)(380\,N)$$
$$= 3040\,J \approx 3.0 \times 10^3\,J$$

16. ●●If it requires $5.0\,J$ of work to stretch a particular spring by $2.0\,cm$ from its equilibrium length, how much more work will be required to stretch it an additional $4.0\,cm$?

17. ●●In following figure assume the distance axis is the x axis and that $a = 10.0\,m$ and $b = 30.0\,m$. Estimate the work done by this force in moving a 3.50-kg object from a to b.

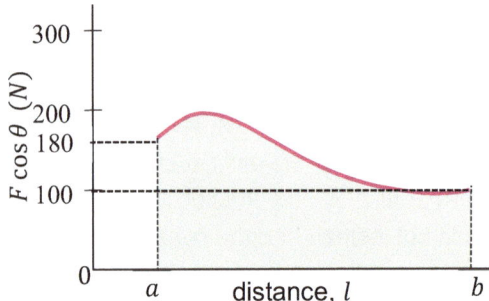

18. ●●You and your bicycle have combined mass $80\,kg$. When you reach the base of a bridge, you are traveling along the road at $5.0\,m/s$. At the top of the bridge, you have climbed a vertical distance of $5.2\,m$ and have slowed to $1.5\,m/s$. You can ignore work done by friction and any inefficiency in the bike or your legs. (a) What is the total work done on you and your bicycle when you go from the base to the top of the bridge? (b) How much work have you done with the force you apply to the pedals?

19. ●●The force on a particle, acting along the x axis, varies as shown in adjoining figure. Determine the work done by this force to move the particle along the x axis: (a) from $x = 0$ to $x = 10.0\,m$; (b) from to $x = 0$ to $x = 15.0\,m$.

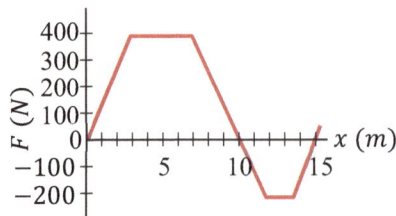

20. ●●The resistance of a packing material to a sharp object penetrating it is a force proportional to the fourth power of the penetration depth x; that is, $\vec{F} = -kx^4\hat{\imath}$. Calculate the work done to force a sharp object a distance d into the material.

21. ●●Assume that a force acting on an object is given by $\vec{F} = ax\hat{\imath} + by\hat{\jmath}$, where the constants $a = 3.0\,Nm^{-1}$ and $b = 4.0 Nm^{-1}$. Determine the work done on the object by this force as it moves in a straight line from the origin to $\vec{r} = (10\hat{\imath} + 20\hat{\jmath})m$.

22. ●●The force exerted on an object is $\vec{F} = F_0(x/x_0 - 1)\hat{\imath}$. Find the work done in moving the object from $x = 0$ to $x = 3x_0$

23. ●●A 2800-kg space vehicle, initially at rest, falls vertically from a height of $3300\,km$ above the Earth's surface. Determine how much work is done by the force of gravity in bringing the vehicle to the Earth's surface (Given that, radius of earth, $R = 6.38 \times 10^6 m$).

24. ●●Two springs, each with force constant k and unstretched length l_0, are connected in a straight line as shown in following figure. (a) Find an expression for the work required to move the point of attachment between the two springs a perpendicular distance x from the equilibrium point. (b) Use the binomial expansion to find the first nonvanishing term in the expression for the work when $x \ll l_0$.

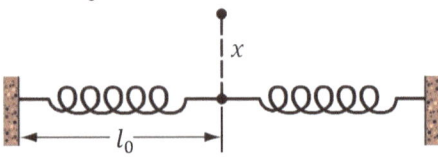

25. ●●●Four springs, each with force constant k and unstretched length l_0, are connected as shown in following figure. The springs obey hooks law for both stretching and compression. Show that the work required to move the point of attachment from the equilibrium position in a straight line to the point

(x, y) (with $x \ll l_0$ and $y \ll l_0$) is $W = kd^2$, where, $d^2 = x^2 + y^2$.

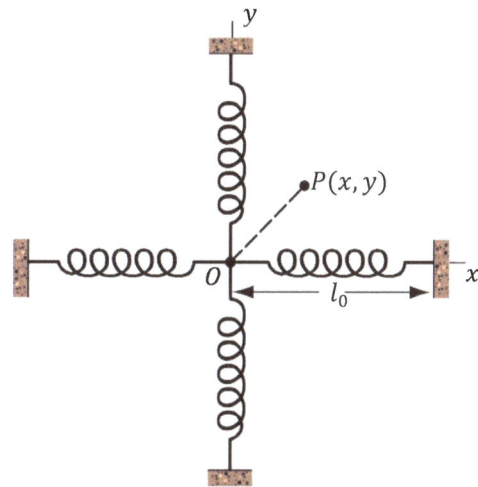

26. ●●●A 3.0-m-long steel chain is stretched out along the top level of a horizontal platform at a construction site, in such a way that $2.0\ m$ of the chain remains on the top level and $1.0\ m$ hangs vertically, (see adjoining figure). At this point, the force on the hanging segment is sufficient to pull the entire chain over the edge. Once the chain is moving, the kinetic friction is so small that it can be neglected. How much work is performed on the chain by the force of gravity as the chain falls from the point where $2.0\ m$ remains on the platform to the point where the entire chain has left the platform? (Assume that the chain has a linear weight density of $18\ N/m$)

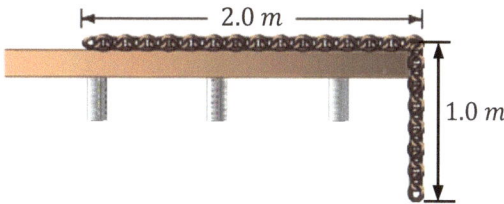

27. ●●● We usually neglect the mass of a spring if it is small compared to the mass attached to it. But in some applications, the mass of the spring must be taken into account. Consider a spring of unstretched length l and mass M_s uniformly distributed along the length of the spring. A mass m is attached to the end of the spring. One end of the spring is fixed and the mass m is allowed to vibrate horizontally without friction (see following figure). Each point on the spring moves with a velocity proportional to the distance from that point to the fixed end. For example, if the mass on the end moves with speed v_0 the midpoint of the spring moves with speed $v_0/2$ Show that the kinetic energy of the mass plus spring system when the mass is moving with velocity v is

$$K = \frac{1}{2}Mv^2$$

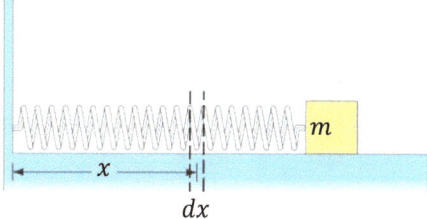

where $M = m + \frac{1}{3}M_s$ is the "effective mass" of the system.
[Hint: Let D be the total length of the stretched spring. Then the velocity of a mass dm of a spring of length dx located at x is $v(x) = v_0(x/D)$. Note also that $dm = dx(M_s/D)$]

28. ●●●A 2.5-kg textbook is forced against a horizontal spring of negligible mass and force constant $250\ N/m$ compressing the spring a distance of $0.25\ m$. When released, the textbook slides on a horizontal tabletop with coefficient of kinetic friction $\mu_k = 0.3$. Use the work–energy theorem to find how far the text book moves from its initial position before coming to rest.

29. ●●A physics professor is pushed up a ramp inclined upward at 30° above the horizontal as he sits in his desk chair that slides on frictionless rollers. The combined mass of the professor and chair is $85\ kg$. He is pushed $2.5\ m$ along the incline by a group of students who together exert a constant horizontal force of 600 N. The professor's speed at the bottom of the ramp is $2\ m/s$. Use the work–energy theorem to find his speed at the top of the ramp

30. ●●A 263-g block is dropped onto a vertical spring with force constant $k = 2.52\ N/cm$ (see following figure). The block sticks to the spring, and the spring compresses $11.8\ cm$ before coming momentarily to rest. While the spring is being compressed, how much work is done (a) by the force of gravity and (b) by the spring? (c) What was the speed of the block just before it hit the spring? (d) If this initial speed of the block is doubled, what is the maximum compression of the spring? Ignore friction.

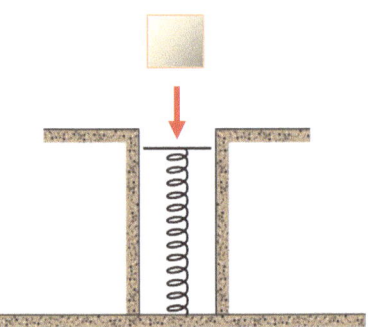

31. ●●An object of mass m accelerates uniformly from rest to a speed v_f in time t_f. (a) Show that the work done on the object as a function of time t, in terms of v_f and t_f, is

$$W = \frac{1}{2}m\frac{v_f^2}{t_f^2}t^2$$

(b) As a function of time t, what is the instantaneous power delivered to the object?

32. ●●Suppose that the blades on a helicopter push vertically down the cylindrical column of air they sweep out as they rotate. The total mass of the helicopter is $1820\ kg$ and the length of the blades is $4.88\ m$. Find the minimum power needed to keep the helicopter airborne. Assume that the density of air is $1.23\ kg/m^3$.

33. ●●2.8-kg block slides over the smooth, icy hill shown in adjoining figure. The top of the hill is horizontal and $70\ m$ higher than its base. What minimum speed must the block have at the base of the hill in order for it to pass over the pit at the far side of the hill?

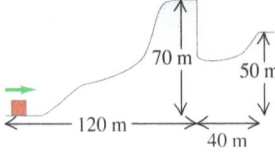

34. ●●A 57-kg woman runs up a flight of stairs having a rise of $4.5\ m$ in $3.5\ s$. What average power must she supply?

35. ●●How much power, in horsepower, must be developed by the engine of a 1600-kg car moving at $26\ m/s$ on a level road if the forces of resistance total $720\ N$ (Given that $1 hp = 745.7 W$)?

36. ●●The motor on a water pump is rated at $6.6\ hp$. From how far down a well can water be pumped up at the rate of $220\ gal/min$? (Given $1\ gallon = 3.785\ litres$)

37. ●●Show that the speed v reached by a car of mass m that is driven with constant power P is given by

$$v = (3xP/m)^{1/3}$$

where x is the distance travelled from rest.

38. ●●(a) Show that the power output of an airplane cruising at constant speed v in level flight is proportional to v^3. Assume that the aerodynamic drag force is given by $D = bv^2$. (b) By what factor must the engines' power be increased to increase the air speed by 25.0%?

39. ●●The power output from a motor on a trolley is a function of velocity and is given by $P(v) = av(b - v^2)$, where a and b are constants and $P = 0$ for $v^2 > b$. (a) At what speed is the maximum power output from the motor? (b) At what speed is maximum force exerted by the motor? (c) At $v = 0$ the power output is zero. Does this mean that the motor will be unable to move the trolley if it is originally at rest? Explain.

40. ●●A particle moves along the x axis under the influence of a conservative force that is described by

$$\vec{F} = -\alpha x e^{-\beta x^2}\hat{\imath}$$

where α and β are constants. Find the potential energy function $U(x)$.

41. ●●A block of mass $m = 5.00\ kg$ is released from point A and slides on the frictionless track shown in following figure. Determine (a) the block's speed at points B and C and (b) the net work done by the gravitational force on the block as it moves from point A to point C.

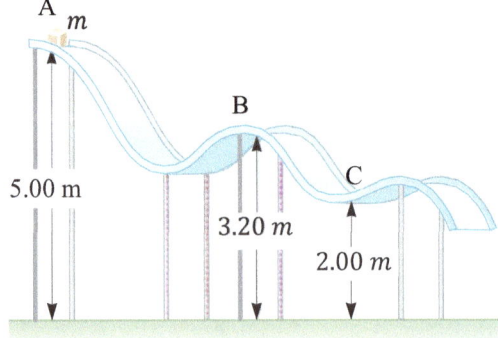

42. ●●A frictionless roller-coaster car starts at point A in following figure with speed v_0. What will be the speed of the car (a) at point B, (b) at point C, and (c) at point D? Assume that the car can be considered a particle and that it always remains on the track.

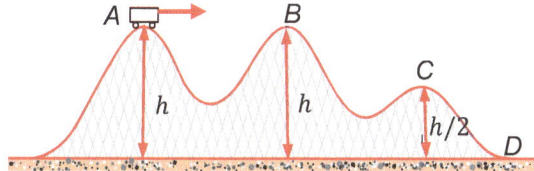

43. ●● The system shown in following figure consists of a light, inextensible cord, light, frictionless pulleys, and blocks of equal mass. Notice that block B is attached to one of the pulleys. The system is initially held at rest so that the blocks are at the same height above the ground. The blocks are then released. Find the speed of block A at the moment the vertical separation of the blocks is h.

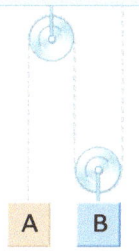

44. ●● Adjoining figure shows a 7.94-kg stone resting on a spring. The spring is compressed 10.2 cm by the stone. (a) Calculate the force constant of the spring. (b) The stone is pushed down an additional 28.6 cm and released. How much potential energy is stored in the spring just before the stone is released? (c) How high above this new (lowest) position will the stone rise?

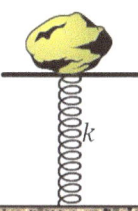

45. ●● A 2.14-kg block is dropped from a height of 43.6 cm onto a spring of force constant $k = 18.6 \, N/cm$, as shown in adjoining figure. Find the maximum distance the spring will be compressed.

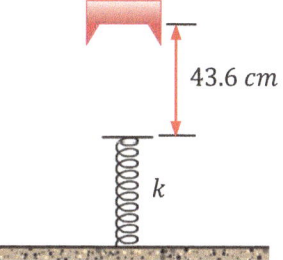

46. ●● As shown in following figure, a green bead of mass 25 g slides along a straight wire. The length of the wire from point A to point B is 0.600 m, and point A is 0.200 m higher than point B. A constant friction force of magnitude 0.025 0 N acts on the bead. (a) If the bead is released from rest at point A, what is its speed at point B? (b) A red bead of mass 25 g slides along a curved wire, subject to a friction force with the same constant magnitude as that on the green bead. If the green and red beads are released simultaneously from rest at point A, which bead reaches point B with a higher speed? Explain.

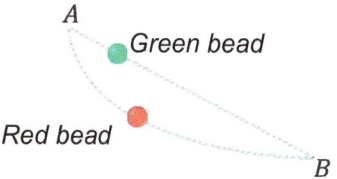

47. ●● A child of mass m starts from rest and slides without friction from a height h along a slide next to a pool (see adjoining figure). She is launched from a height $h/5$ into the air over the pool. We wish to find the maximum height she reaches above the water in her projectile motion. (a) Is the child–Earth system isolated or non-isolated? Why? (b) Is there a nonconservative force acting within the system? (c) Define the configuration of the system when the child is at the water level as having zero gravitational potential energy. Express the total energy of the system when the child is at the top of the waterslide. (d) Express the total energy of the system when the child is at the launching point. (e) Express the total energy of the system when the child is at the highest point in her projectile motion. (f) From parts (c) and (d), determine her initial speed vi at the launch point in terms of g and h. (g) From parts (d), (e), and (f), determine her maximum airborne height y_{max} in terms of h and the launch angle θ. (h) Would your answers be the same if the waterslide were not frictionless? Explain.

48. ●● The spring in adjoining figure has a spring constant of 1000 N/m. It is compressed 15 cm, then launches a 200 gram block. The horizontal surface is frictionless, but the block's coefficient of kinetic friction on the incline is 0.20. What distance d does the block sail through the air?

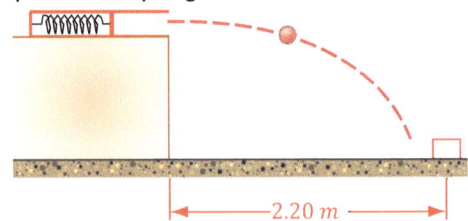

49. •• Two children are playing a game in which they try to hit a small box on the floor with a marble fired from a spring-loaded gun that is mounted on a table. The target box is 2.20 m horizontally away from the edge of the table (see following figure). Manoj compresses the spring 1.10 cm, but the marble falls 27.0 cm short. How far should Tony compress the spring to score a hit?

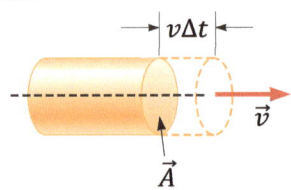

50. ••• As it plows a parking lot, a snowplow pushes an ever-growing pile of snow in front of it. Suppose a car moving through the air is similarly modelled as a cylinder of area A pushing a growing disk of air in front of it. The originally stationary air is set into motion at the constant speed v of the cylinder as shown in following figure. In a time interval Δt, a new disk of air of mass Δm must be moved a distance $v\Delta t$ and hence must be given a kinetic energy $\frac{1}{2}(\Delta m)v^2$. Using this model, show that the car's power loss owing to air resistance is $\frac{1}{2}\rho A v^3$ and that the resistive force acting on the car is $\frac{1}{2}\rho A v^2$, where ρ is the density of air. Compare this result with the empirical expression $\frac{1}{2}D\rho A v^2$ for the resistive force.

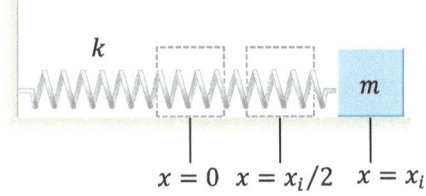

51. ••• A horizontal spring attached to a wall has a force constant of $k = 850 \ N/m$. A block of mass $m = 1.00 \ kg$ is attached to the spring and rests on a frictionless, horizontal surface as in adjoining figure. The block is pulled to a position $x_i =$ 6.00 cm from equilibrium and released. Find the elastic potential energy stored in the spring when the block is 6.00 cm from equilibrium and when the block passes through equilibrium. (b) Find the speed of the block as it passes through the equilibrium point. (c) What is the speed of the block when it is at a position $x_i/2 = 3.00 \ cm$? (d) Why isn't the answer to part (c) half the answer to part (b)?

52. ••• A uniform chain of length 8.00 m initially lies stretched out on a horizontal table. (a) Assuming the coefficient of static friction between chain and table is 0.6, show that the chain will begin to slide off the table if at least 3.0 m of it hangs over the edge of the table. (b) Determine the speed of the chain as its last link leaves the table, given that the coefficient of kinetic friction between the chain and the table is 0.4.

53. ••• The so-called Yukawa potential energy

$$U(r) = -\frac{r_0}{r}U_0 e^{-r/r_0}$$

gives a fairly accurate description of the interaction between nucleons (i.e., neutrons and protons, the constituents of the nucleus). The constant r_0 is about 1.5×10^{-15} m and the constant U_0 is about 50 MeV. (a) Find the corresponding expression for the force of attraction. (b) To show the short range of this force, compute the ratio of the force at $r = 2r_0$, $4r_0$, and $10r_0$ to the force at $r = r_0$.

32.3. MULTIPLE CHOICE PROBLEMS

1. When a conservative force does positive work on a body, then
   (A) its potential energy must decrease.
   (B) its potential energy must increase.
   (C) its total energy must decrease
   (D) its kinetic energy must increase.

2. A block of mass 10 kg accelerates uniformly from rest to a speed of 2 m/s in 20 sec. The average power developed in time interval of 0 to 20 sec is
   (A) 10W   (B) 1W

(C) 20W (D) 2W

3. An object is attached to a vertical spring and is allowed to fall under the gravity. What is the distance traversed by the object before being stopped?
(A) mg/k (B) $2mg/k$
(C) $mg/2k$ (D) none of these.

4. When a man walks on a horizontal surface with constant velocity work done by the
(A) frictional force is zero
(B) contact force is zero
(C) gravity is zero
(D) man is zero

5. A block of mass m is initially at rest on a horizontal surface. A constant horizontal force of magnitude $mg/2$ is applied to the box directed to the right. The coefficient of friction of the surface changes with the distance pushed as $\mu = \mu_0 x$ where x is the distance from the initial location. For what distance is the box pushed until it comes to rest again?

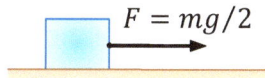

(A) $\dfrac{1}{4\mu_0}$ (B) $\dfrac{1}{2\mu_0}$ (C) $\dfrac{1}{\mu_0}$ (D) $\dfrac{2}{\mu_0}$

6. A pumping machine pumps a liquid at a rate of 60 cc per minute at a pressure of 1.5 atm. The power delivered by the machine is
(A) 9 watt (B) 6 watt
(C) 9 kW (D) None of these

7. A block of mass m moving on a smooth horizontal floor, with a constant velocity v_0 collides with a light spring of stiffness constant k which is rigidly fixed horizontally with a vertical wall. If the maximum force imparted by the spring on the block is F, then:
(A) $F \propto \sqrt{m}$ (B) $F \propto \sqrt{k}$
(C) $F \propto v_0$ (D) all of these

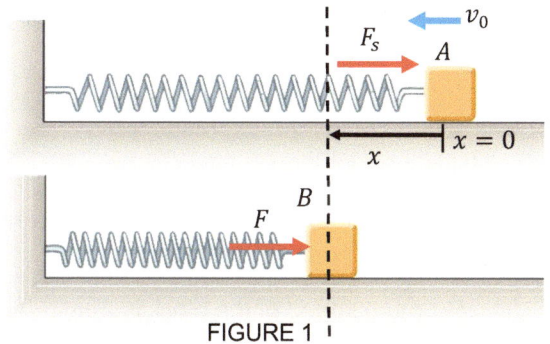

FIGURE 1

8. A block of mass m is moving with a constant acceleration a on a frictional plane. If the coefficient of friction between the block and ground is μ, the power delivered by the external agent after a time t from the beginning is equal to:
(A) $ma^2 t$ (B) $\mu mgat$
(C) $\mu m(a + \mu g)gt$ (D) $m(a + \mu g)at$

9. The ratio of work done by gravity on the block in the two conditions shown in following figure when it reaches to ground from height h, is-

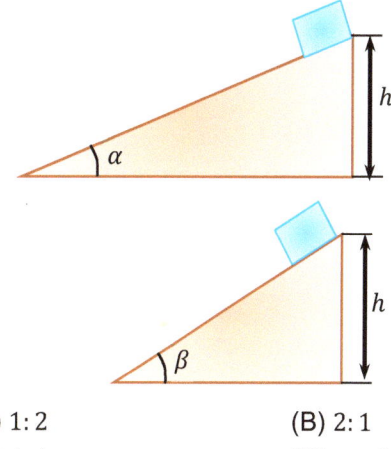

(A) $1:2$ (B) $2:1$
(C) $1:1$ (D) $\alpha : \beta$

10. A spring, placed horizontally on a rough surface is compressed against a block of mass m, placed on the same surface so as to store maximum energy in the spring. If the coefficient of friction between the block and the surface is μ, the potential energy stored in the spring is

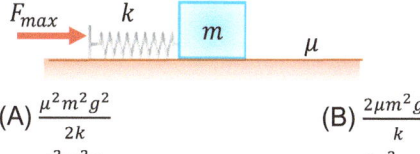

(A) $\dfrac{\mu^2 m^2 g^2}{2k}$ (B) $\dfrac{2\mu m^2 g^2}{k}$
(C) $\dfrac{\mu^2 m^2 g}{2k}$ (D) $\dfrac{3\mu^2 mg^2}{k}$

11. An object of mass m is tied to a string of length L and a variable horizontal force is applied on it which starts with zero magnitude and gradually increases until the string makes an angle θ with the vertical. Work done by the variable force F is:

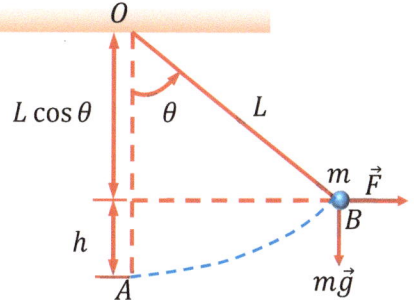

(A) $mgL(1-\sin\theta)$ (B) mgL
(C) $mgL(1-\cos\theta)$ (D) $mgL(1+\cos\theta)$

12. When a body of mass M slides down an inclined plane of inclination θ, having coefficient of friction μ through a distance s, the work done against friction is
(A) $\mu Mg\cos\theta\, s$ (B) $\mu Mg\sin\theta\, s$
(C) $Mg(\mu\cos\theta-\sin\theta)s$ (D) None of the above

13. An object is pushed and pulled along a certain angle with respect to horizontal and is moved the same distance on the ground. The work done by frictional force will have

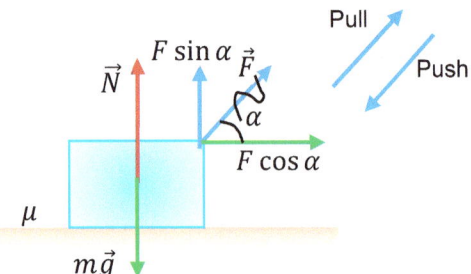

(A) push effect (B) pull effect
(C) remain same (D) data insufficient

14. A body is acted upon by a force which is inversely proportional to the distance covered. The work done will be proportional to:
(A) s (B) s^2
(C) $\sqrt{s}$ (D) None of the above

15. Find the horizontal velocity of the block when it reaches the point Q. Assume the surface to be frictionless. Take $g = 9.8\, m/s^2$.

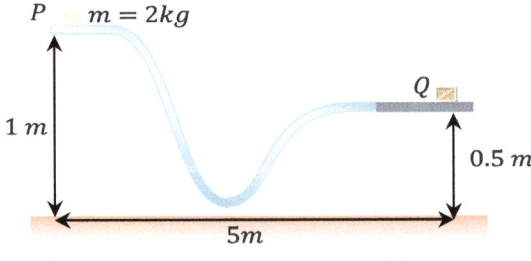

(A) 4 m/s (B) 5 m/s
(C) 3.13 m/s (D) 3.6 m/s

16. A body of mass **5 kg** initially at rest, is subjected to a force of **40 N**. Find the kinetic energy acquired by the body at the end of **5** seconds
(A) 4000 J (B) 6000 J
(C) 6249 J (D) 6145 N

17. A body is constrained to move along z-axis of the co-ordinate system is being applied by a constant force $\vec{F} = (2\hat{i}+3\hat{j}+4\hat{k})$. Find the work done by this force in moving the body over a distance of 5m along z-axis.
(A) 30 J (B) 20 J
(C) 40 J (D) 61 J

18. A body moves from point A to point B under the action of a force varying in magnitude as shown in the force displacement graph. Find total work done by the force

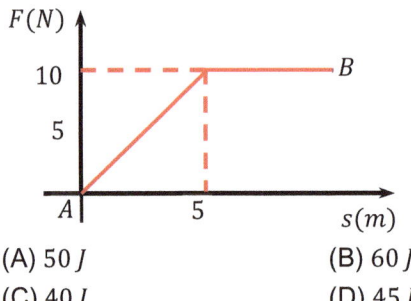

(A) 50 J (B) 60 J
(C) 40 J (D) 45 J

19. A body dropped form a height 'H' reaches the ground with a speed of $1.1\sqrt{gH}$. Calculate the work done by air friction?
(A) $0.395\, mgH$ (B) $-0.395\, mgH$
(C) $0.400\, mgH$ (D) $-0.400\, mgH$

20. A block of mass $m = 2kg$ is attached to two unstretched springs of force constants $k_1 = 100\, N/m$ and $k_2 = 125\, N/m$ respectively. The block is displaced towards left through a distance of 10 cm and released. Find the speed of the block as it passes through the mean position

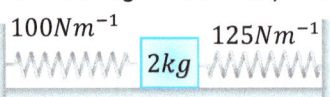

(A) 1.06 m/s (B) 1.02 m/s
(C) 1.04 m/s (D) 1.05 m/s

21. A spring block system is placed on a rough horizontal surface having coefficient of friction μ. The spring is given initial elongation $\frac{3\mu mg}{k}$ and the block is released from rest. For the subsequent motion

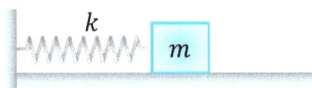

(A) Initial acceleration of block is $2\mu g$.
(B) Maximum compression in spring is $\frac{\mu mg}{k}$
(C) Minimum compression in spring is zero.
(D) Maximum speed of the block is $2\mu g\sqrt{\frac{m}{k}}$

22. A 16 kg block moving on a frictionless horizontal surface with a velocity of $4 m/s$ compresses an ideal spring and comes to rest. If the force constant of the spring is $100 \, N/m$, then how much is the spring compressed?
(A) 1.7 m (B) 1.6 m (C) 1.4 m (D) 1.8 m

23. A small block of mass $500 \, gm$ is pressed against a horizontal spring fixed at one end to a distance of $10 \, cm$. when released, the block moves horizontally till it leaves the spring. Where it will hit the ground $5 \, m$ below the spring. (Take $k = 100 \, N/m$ of spring, $g = 10 \, m/s^2$)

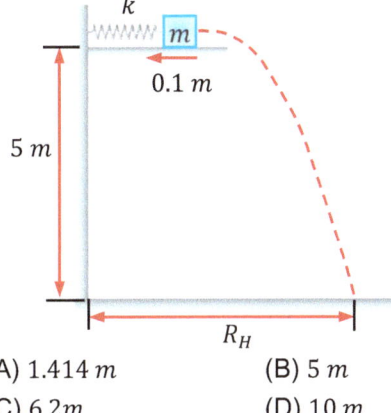

(A) $1.414 \, m$ (B) $5 \, m$
(C) $6.2 m$ (D) $10 \, m$

24. A block of mass m moving with a velocity v_0 on a smooth horizontal floor collides with a light spring of stiffness k that is rigidly fixed horizontally with a vertical wall. If the maximum force imparted by the spring on the block is F, then the value of F is-
(A) $\sqrt{km}v_0$ (B) $4\sqrt{km}v_0$
(C) $2\sqrt{km}v_0$ (D) $\sqrt{km}$

25. A bus of mass m produces a constant power P. If the resistance to motion is R. Find the maximum speed at which the bus can travel on level road and acceleration of bus when it is travelling at half of its maximum speed.
(A) $a = R$ (B) $a = \frac{R}{m}$
(C) $a = \frac{R}{2m}$ (D) $a = \frac{1}{m}$

26. A uniform rod of length L and mass m hinged at one end is hanging vertically. The other end is now raised until it makes an angle $60°$ with vertical. How much work is required?

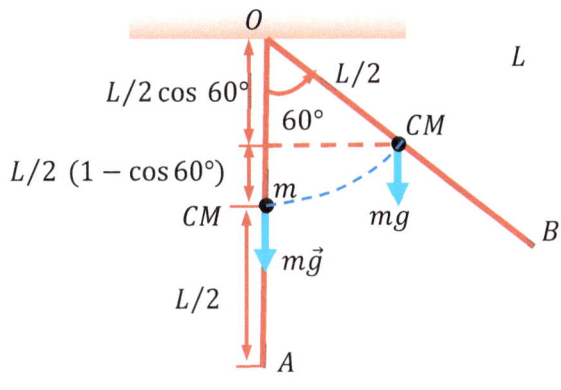

(A) mgL (B) $\frac{mgL}{8}$
(C) $\frac{mgL}{6}$ (D) $\frac{mgL}{4}$

(31-35) A block of ice mass $10 \, kg$ slides down an incline $5 \, m$ long and $3 \, m$ high. A man pushes up on the ice block parallel to the incline so that it slides down at constant speed. The coefficient of friction between the ice and the incline is 0.1 and $g = 10 \, m/s^2$. Find:

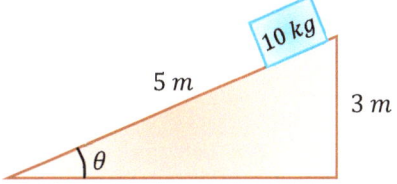

27. The work done by the man on the block
(A) $W_m = -300 \, J$ (B) $W_m = -260 \, J$
(C) $W_m = -400 \, J$ (D) $W_m = -560 \, J$

28. The work done by gravity on the block
(A) $W_{gravity} = 300 \, J$ (B) $W_{gravity} = 400 \, J$
(C) $W_{gravity} = 600 \, J$ (D) $W_{gravity} = 500 \, J$

29. The work done by the surface on the block
(A) $W_{surface} = +40 \, J$ (B) $W_{surface} = -42 \, J$
(C) $W_{surface} = -41 \, J$ (D) $W_{surface} = -40 \, J$

30. The work done by the resultant forces on the block
(A) $W = 0 J$ (B) $W = 2J$
(C) $W = 3J$ (D) $W = 6J$

31. The change in K.E. of the block-
(A) $KE = 0$ (B) $KE = 6$
(C) $KE = 4$ (D) $KE = 3$

32. If a constant force $\vec{F} = 3\hat{\imath} + 4\hat{\jmath} + 5\hat{k}$ is acting on a particle and displacement of particle becomes $\vec{r} = 3\hat{\imath} + 4\hat{\jmath} + 6\hat{k}$. Find net work done by force $\vec{F}$.
(A) $55 \, J$ (B) $80 \, J$
(C) $90 \, J$ (D) $100 \, J$

33. A projectile is projected with initial velocity 20 m/s and at an angle of 30° from horizontal. Find the total work done when it will hit the ground.
   (A) 0   (B) 10   (C) 20   (D) 30

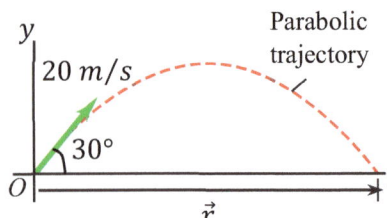

34. A particle of mass $1\ kg$ is moving along x-axis and a force F is also acting along x-axis in such a way that its displacement is varying as: - $x = 3t^2$. The work done by force $\vec{F}$, when it moves $2m$, is-.
   (A) 12 J   (B) 16 J   (C) 32 J   (D) 42 J

35. If a projectile is projected with initial speed u and angle θ from horizontal then, what will be its average power up to time when it will hit the ground again.
   (A) 100   (B) 200   (C) 300   (D) 0

36. An 8 kg block accelerates uniformly from rest to a velocity of $4\ ms^{-1}$ in 40 second. The instantaneous power at the end of 8 second is
   (A) 0.64 W   (B) 0.32 W
   (C) 0.16 W   (D) 0.08 W.

37. An engine is working at a constant power draws a load of mass m against a resistance r. Find maximum speed of load and time taken to attain half this speed.
   (A) $t = \frac{Pm}{8r^2}$ (B) $t = \frac{Pm}{8r}$
   (C) $t = \frac{Pm}{r^2}$ (D) $t = \frac{Pm}{9r^2}$

38. Block m has given a velocity v_0 towards right. If the displacement of the block at any time t is given by, $x = A \sin \omega t$ (A is known as amplitude and ω is angular velocity), then the maximum power is delivered by the spring force during the entire motion at the time-

   (A) $\frac{5\pi}{4\omega}$ (B) $\frac{3\pi}{7\omega}$ (C) $\frac{3\pi}{4\omega}$ (D) $\frac{1\pi}{4\omega}$

39. A block of mass 2 kg is dragged by a force of 20 N on a smooth horizontal surface. It is observed from three reference frames ground, observer A and observer B. Observer A is moving with constant velocity of $10\ m/s$ and B is moving with constant acceleration of $10\ m/s^2$. The observer B and block starts simultaneously at $t = 0$.

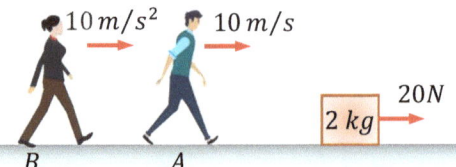

   Column I   Column II
   (A) Work energy theorem is applicable in   (P) 100 J
   (B) Work done on block in 1s as observed by ground is   (Q) −100 J
   (C) Work done on block is 1 s as observed by observer A is   (R) zero
   (D) Work done on block in 1 s as observed by observer B is   (S) only ground & A
                                                 (T) all frames ground, A & B

40. If a spring mass system is placed on a horizontal smooth surface. Then what will be the net work done by spring force after one time period.
   (A) $\frac{1}{2}km^2$ (B) $-\frac{1}{2}km^2$
   (C) $-\frac{1}{4}km^2$ (D) $+\frac{1}{4}km^2$

41. The force $\vec{F} = x\hat{i} + y\hat{j} + z\hat{k}$ is-
   (A) conservative
   (B) non conservative
   (C) variable
   (D) none of these.

42. A particle of mass m is moving in a circular path of constant radius r such that centripetal acceleration a_c is varring with time t as $a_c = k^2 r t^2$ where k is constant. The power delivered to the particle by the forces acting on it.
   (A) $\frac{mk^4 r^2 t^5}{4}$ (B) $\frac{mk^4 r^2 t^5}{5}$
   (C) $\frac{mk^4 r^2 t^5}{3}$ (D) $\frac{mk^4 r^2 t^5}{9}$

43. The total compression in the spring is-

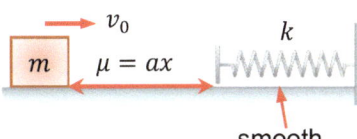

(A) $x = \sqrt{\frac{mv_0^2 - amg}{k}}^2$

(B) $x = \frac{mv_0^2 - amgL^2}{k}$

(C) $x = \sqrt{\frac{mv_0^2 - amgL^2}{k^2}}$

(D) $\Rightarrow x = \sqrt{\frac{mv_0^2 - amgL^2}{\frac{k}{2}}}$

(44-46) A particle of mass $m = 1\,kg$ is moving along x-axis and a single conservative force $F(x)$ acts on it. The potential energy of particle is given by $U(x) = (x^2 - 6x + 14)\,J$ where x is in meters. At $x = 3\,m$ the particle has kinetic energy of $15\,J$.

44. The total mechanical energy of the particle is
    (A) 15 J (B) 5 J
    (C) 20 J (D) can't be determined

45. The maximum speed of the particle is
    (A) $5\,m/s$ (B) $\sqrt{30}\,m/s$
    (C) $\sqrt{40}\,m/s$ (D) $\sqrt{10}\,m/s$

46. The largest value of x (position of particle) is
    (A) $3 + \sqrt{5}$ (B) $3 - \sqrt{5}$
    (A) $3 + \sqrt{15}$ (B) $6 + \sqrt{15}$

47. Power applied to a particle varies with time as $P = (4t^3 - 5t + 2)$ watt, where t is in second. Find the change is its $K.E.$ between time $t = 2$ and $t = 4$ sec.
    (A) 212 J (B) 213 J
    (C) 214 J (D) 215 J

48. An engine develops $10\,kW$ of power. The time taken by it in lifting a mass of $200\,kg$ to a height $40\,m$. ($g = 10\,ms^{-2}$)
    (A) 4 s (B) 5 s
    (C) 8 s (D) 10 s

49. The work done by a force, when a body get displaced perpendicular to the applied force.
    (A) zero (B) maximum
    (C) minimum (D) None of these.

50. A tube-well pump out $2400\,kg$ of water per minute. If water is coming out with a velocity of $3m/s$, the power of the pump is
    (A) 120W (B) 180 W
    (C) 240 W (D) 90 W

32.4. MULTIPLE CHOICE ASSIGNMENTS

32.4.1. LEVEL 1

32.4.1.1. WORK

1. A man pushes a wall and fails to displace it. He does –
   (A) negative work
   (B) positive but not maximum work
   (C) no work at all
   (D) maximum work

2. Work done in time t on a body of mass m which is accelerated from rest to a speed v in time t_1 as a function of time t is given by-
   (A) $\frac{1}{2}m\frac{v}{t_1}t^2$ (B) $m\frac{v}{t_1}t^2$
   (C) $\frac{1}{2}m\left(\frac{mv}{t_1}\right)^2 t^2$ (D) $\frac{1}{2}m\frac{v^2}{t_1^2}t^2$

3. A particle moves under the effect of a force $F = cx$ from $x = 0$ to $x = x_1$. The work done in the process is-
   (A) cx_1 (B) cx_1^2
   (C) cx_1^3 (D) zero.

4. A body travels through a distance of $10\,m$ on a straight line, under the influence of $5\,N$. If the work done by the force is $25J$, the angle between the force and displacement is-
   (A) 0º (B) 30º (C) 60º (D) 90º

5. A block is moved from rest through a distance of $4m$ along a straight line path. The mass of the blocks is $5\,kg$. and the force acting on it is $20\,N$. If the kinetic energy acquired by the block be $40J$, at what angle to the path the force is acting-
   (A) 30º (B) 60º
   (C) 45º (D) none of the above

6. The work done in pushing a block of mass 10 kg from bottom to the top of a frictionless inclined plane 5 m long and 3 m high is ($g = 9.8\,m/s^2$)-
   (A) 392 J (B) 294 J
   (C) 98 J (D) 0.98 J

7. The figure shows the force (F) versus displacement (s) graph for a particle of mass $m = 2kg$ initially at rest

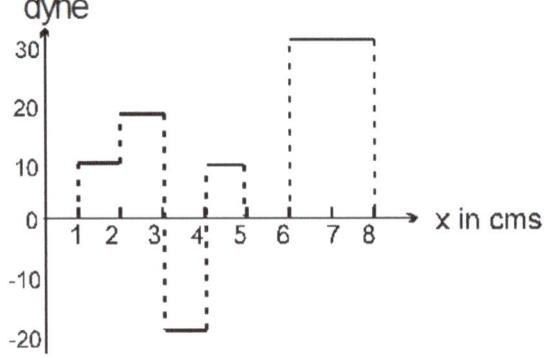

(i) The maximum speed of the particle occurs at $x = \ldots m$

(ii) The maximum speed of the particle is $\ldots ms^{-1}$

(iii) The particle once again has its speed zero at $x = \ldots m$

(A) 5, 3, 6
(B) 3, 4.18, 6
(C) 6, 5, 3
(D) 4, 5, 6

8. A force $\vec{F} = (3x\hat{i} + 4\hat{j})$ newton (where x is in metres) acts on a particle which moves from a position $(2m, 3m)$ to $(3m, 0m)$. Then the work done by this force is-
(A) 7.5 J
(B) −12 J
(C) − 4.5 J
(D) +4.5 J

9. A chain of mass M is placed on a smooth table with $1/n$ of its length L hanging over the edge. The work done in pulling the hanging portion of the chain back to the surface of the table is –
(A) MgL/n
(B) $MgL/2n$
(C) MgL/n^2
(D) $MgL/2n^2$

10. The relationship between force and position is shown in adjoining figure (in one dimensional case). The work done in displacing a body from $x = 1cm$ to $x = 5\,cm$ is:

(A) 20 erg
(B) 60 erg
(C) 70 erg
(D) 700 erg

11. A force $\vec{F} = 2\hat{i} - 3\hat{j} + 7\hat{k}$ (N) acts on a particle which undergoes a displacement $\vec{r} = 7\hat{i} + 3\hat{j} - 2\hat{k}$ (m). Calculate the work done by the force
(A) 37 J
(B) − 9 J
(C) 49 J
(D) 14 J

32.4.1.2. POWER

12. A pump ejects $12000\,kg$ of water at speed of $4\,m/s$ in 40 second. Find the average rate at which the pump is working
(A) 0.24 KW
(B) 2.4 W
(C) 2.4 KW
(D) 24 W

13. An object of mass m accelerates uniformly from rest to a speed v_f in time t_f. Then the instantaneous power delivered to the object, as a function of time t is –
(A) $mt\left(\dfrac{v_f}{t_f}\right)^2$
(B) $mt\dfrac{v_f}{t_f}$
(C) $\dfrac{1}{2}mt^2\left(\dfrac{v_f}{t_f}\right)^2$
(D) $\dfrac{1}{2}mt^2\left(\dfrac{v_f}{t_f}\right)$

14. A self-propelled vehicle of mass m whose engine delivers constant power P has an acceleration $a = \dfrac{P}{mv}$ (assume that there is no friction). In order to increase its velocity from v_1 to v_2, the distance it has to travel will be
(A) $\dfrac{3P}{m}(v_2^2 - v_1^2)$
(B) $\dfrac{m}{3P}(v_2^3 - v_1^3)$
(C) $\dfrac{m}{3P}(v_2^2 - v_1^2)$
(D) $\dfrac{m}{3P}(v_2 - v_1)$

[Hint: Use $a = v.\dfrac{dv}{dx}$, i.e. $a\,dx = v\,dv$

or $\dfrac{P}{mv}dx = v\,dv$

$\Rightarrow dx = \dfrac{m}{P}v^2 dv$ or $\int_0^x dx = \dfrac{m}{P}\int_{v_1}^{v_2} v^2 dv$

or $x = \dfrac{m}{3P}(v_2^3 - v_1^3)$]

15. If a force F is applied on a body and it moves with a velocity v, the power will be-
(A) $F\,v$
(B) F / v
(C) F/v^2
(D) Fv^2

16. A body of mass m accelerates uniformly form rest to v_1 in time t_1. As a function of t, the instantaneous power delivered to the body is-
(A) $\dfrac{mv_1 t}{t_1}$
(B) $\dfrac{mv_1^2 t}{t_1}$
(C) $\dfrac{mv_1 t^2}{t_1}$
(D) $\dfrac{mv_1^2 t}{t_1^2}$

17. A body is moved along a straight line by a machine delivering constant power. The distance moved by the body in time t is proportional to-

(A) $t^{1/2}$ (B) $t^{3/4}$
(C) $t^{3/2}$ (D) t^2

32.4.1.3. KINETIC ENERGY

18. A light and a heavy body have equal momentum. Which one has greater $K.E.$?
    (A) the light body
    (B) both have equal $K.E.$
    (C) the heavy body
    (D) data given is incomplete

19. If a man increases his speed by $2\ m/s$, his K.E. is doubled. The original speed of the man is-
    (A) $(2 + \sqrt{2})\ m/s$ (B) $(2 + 2\sqrt{2})\ m/s$
    (C) $4\ m/s$ (D) $(1 + \sqrt{2})\ m/s$

20. A 300 g mass has a velocity of $(3\hat{\imath} + 4\hat{\jmath})\ m/s$ at a certain instant what is its $K.E.$?
    (A) $1.35\ J$ (B) $2.4\ J$
    (C) $3.75\ J$ (D) $7.35\ J$

21. Two bodies of mass $1\ kg$ and $4\ kg$ are moving with equal kinetic energies. The ratio of their linear momentum is-
    (A) $1:2$ (B) $2:1$
    (C) $4:1$ (D) $1:4$

22. The momentum of a body is increased by 50%. The K.E. of the body will be increased by-
    (A) 50 % (B) 125 %
    (C) 330 % (D) 400 %

32.4.1.4. POTENTIAL ENERGY

23. If the unit of force and length be each increased by four times, then the unit of energy is increased by-
    (A) 16 times (B) 8 times
    (C) 2 times (D) 4 times

24. Two springs A and B ($k_A = 2k_B$) are stretched by applying forces of equal magnitudes at the four ends. If the energy stored in A is E, that in B is
    (A) E/2 (B) 2E (C) E (D) E/4

25. _____ of a two-particle system depends only on the separation between the two particles. The most appropriate choice for the blank space in the above sentence is
    (A) kinetic energy
    (B) total mechanical energy
    (C) potential energy
    (D) total energy

32.4.1.5. CONSERVATION OF MECHANICAL ENERGY

26. The principle of conservation of energy implies that-
    (A) the total mechanical energy is conserved.
    (B) the total kinetic energy is conserved
    (C) the total potential energy is conserved.
    (D) sum of all types of energies is conserved.

27. There will be an increase in potential energy of the system if work is done upon the system by-
    (A) any conservation or non-conservation forces
    (B) a non-conservative force
    (C) a conservative force
    (D) none of the above

28. A body of mass 2 kg fall vertically, passing through two points A and B. The speeds of the body as it passes A and B are 1 m/s and 4m/s respectively. The resistance against which the body falls is $9.6N$. What is the distance AB?
    (A) 2m (B) 3m
    (C) 6m (D) 1.5 m

29. The work done by the external forces on a system equals the change in
    (A) total energy (B) kinetic energy
    (C) potential energy (D) none of these

30. The work done by all the forces (external and internal) on a system equals the change in
    (A) total energy (B) kinetic energy
    (C) potential energy (D) none of these

31. A ball of mass m is dropped from a height h on a platform fixed at the top of a vertical spring. The platform is displaced by a distance x. The spring constant is

    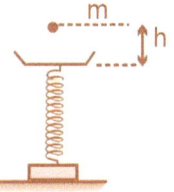

    (A) $\frac{2mg}{x}$ (B) $\frac{2mgh}{x^2}$
    (C) $\frac{2mg(h+x)}{x^2}$ (D) $\frac{2mg(h+x)}{h^2}$

32. A force 'F' stops a body of mass 'm' moving with a velocity 'u' in a distance 's'. The force required to stop a body of double the mass moving with double the velocity in the same distance is-
    (A) 2F (B) 4F

(C) 6F (D) 8F

32.4.2. LEVEL 2

1. A running man has half the kinetic energy that a boy of half his mass has. The man speeds up by 1.0 meter/sec and then has the same kinetic energy as the boy. What were the original speeds of man and boy?
   (A) $2.4\ m/sec,\ 4.8m/sec$
   (B) $4.8m/sec,\ 2.4m/sec$
   (C) $4.2\frac{m}{sec}.\ 4m/sec$
   (D) $\frac{8.4m}{sec}.\ 2m/sec$

2. A block of mass 2kg slipped up a slant plane requires 300J of work. If height of slant is 10m the work done against friction is –
   (A) $100J$ (B) $200J$
   (C) $300J$ (D) zero

3. A chain of mass m and length l is placed on a table with one-sixth of it hanging freely from the table edge. The amount of work. done to pull the chain on the table is
   (A) $mgl/4$ (B) $mgl/6$
   (C) $mgl/72$ (D) $mgl/36$

4. The force required to row a boat over the sea is proportional to the speed of the boat. It is found that it takes $24\ h.p.$ to row a certain boat at a speed of $8km/hr$, the horse power required when speed is doubled –
   (A) $12\ hp.$ (B) $6\ hp.$
   (C) $48\ hp.$ (D) $96hp.$

5. A $50\ kg$ girl is swinging on a swing from rest. Then the power delivered when moving with a velocity of $2m/s$ upwards in a direction making an angle $60°$ with the vertical is
   (A) $980W$ (B) $490W$
   (C) $490W$ (D) $245W$

6. A man cycles up a hill rising 1 meter vertically for every 50 metres along the slope. Find the power of the man, if he cycles up at the rate of $3.6\ km/hr$. The weight of the cycle and man is equal to $120\ kg$. Neglect force of friction.
   (A) $32.25\ watt$ (B) $23.52\ watt$
   (C) $25.32\ watt$ (D) $52.32\ watt$

7. From a waterfall, water is pouring down at the rate of 100 kg per second on the blades of turbine. If the height of the fall is $100\ m$, the power delivered to the turbine is approximately equal to
   (A) $100\ kW$ (B) $10kW$
   (C) $1\ kW$ (D) $100\ W$

8. Under the action of a force a $2\ kg$ mass moves such that its position x as a function of time is given by $x = t^3/3$ where x is in meters and t in seconds. The work done by the force in first two seconds is
   (A) 1600 joules (B) 160 joules
   (C) 16 joules (D) 1.6 joules

9. A locomotive of mass m starts moving so that its velocity varies according to the law $v = ks$ where k is constant and s is the distance covered. Find the total work performed by all the forces which are acting on the locomotive during the first t seconds after the beginning of motion.
   (A) $W = mk^4t^2$. (B) $W = m^2k^4t^3$
   (C) $W = mk^4\ t^4$ (D) $W = mk^3t^4$

10. A 242×10^4 kg freight car moving along a horizontal rail road spur track at $7.2\ km/hour$ strikes a bumper whose coil springs experiences a maximum compression of $30\ cm$ in stopping the car. The elastic potential energy of the springs at the instant when they are compressed $15\ cm$ is
    (A) 12.1×10^4 joules (B) 121×10^4 joules
    (C) 1.21×10^4 joules (D) 1.21×10^6 joules

11. A spring is held compressed. Its stored energy is 2.4 joule. Its ends are in contact with masses of 1gm and $48\ gm$ placed on a smooth horizontal surface. When the spring is released, the mass of $48\ gm$ will acquire a velocity of-
    (A) $2.4/49\ ms^{-1}$ (B) $2.4 \times 48/49\ ms^{-1}$
    (C) $10/7\ ms^{-1}$ (D) $10/7\ cms^{-1}$

12. A block of mass m slips down an inclined plane, from height 'h', as shown in the figure. When it reaches the bottom, it presses the spring by a length (spring length $<< h$ and spring constant = k)-

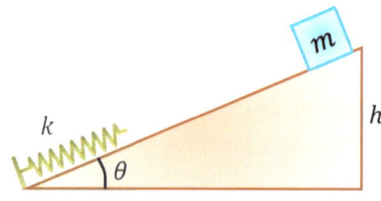

   (A) $(2mgh/k)^{1/2}$ (B) $(mgh/k)^{1/2}$
   (C) $(2gh/mk)^{1/2}$ (D) $(gh/mk)^{1/2}$

13. For the system shown in the adjoining figure, initially the spring is compressed by a distance 'a' from its natural length and when released, it moves to a distance 'b' from its equilibrium position. The decrease in amplitude for half cycle (−a to +b) is:

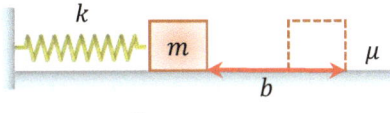

(A) $\frac{\mu mg}{K}$ (B) $\frac{2\mu mg}{K}$ (C) $\frac{\mu g}{K}$ (D) $\frac{K}{\mu mg}$

14. An object of mass m slides down a hill of height h and of arbitrary shape and stops at the bottom because of friction. The coefficient of friction may be different for different segments of the path. Work required to return the object to its position along the same path by a tangential force is

(A) mgh (B) $2mgh$
(C) $-mgh$ (D) it cannot be calculated

15. A 5 kg block is lifted vertically through a height of 5 metre by a force of 60N. Determine (i) the work done by applied force in lifting the block, (ii) the potential energy of the block at 5m, (iii) the kinetic energy of the block at 5 m (iv) the velocity of the block at 5 m-

(A) $300\,J, 245\,J, 55J, 4.69\,m/s$
(B) $200\,J, 245\,J, 50J, 4.69\,m/s$
(C) $150\,J, 150\,J, 50J, 4.69\,m/s$
(D) $300\,J, 245\,J, 100J, 10.69\,m/s$

16. A light rod of length l is pivoted at the upper end. Two masses (each m), are attached to the rod, one at the middle and the other at the free end. What horizontal velocity must be imparted to the lower end mass, so that the rod may just take up the horizontal position?

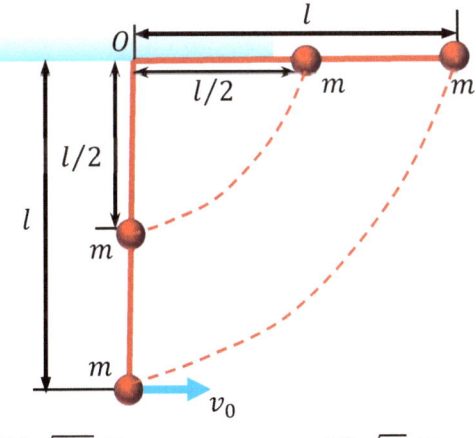

(A) $\sqrt{6lg}/5$ (B) $\sqrt{lg}/5$
(C) $\sqrt{12lg/5}$ (D) $\sqrt{2lg}/5$

17. A machine, which is 72 percent efficient, uses 36 joules of energy in lifting up 1 kg mass through a certain distance. The mass is the allowed to fall through that distance. The velocity at the end of its fall is

(A) $6.6\,ms^{-1}$ (B) $7.2\,ms^{-1}$
(C) $8.1\,ms^{-1}$ (D) $9.2\,ms^{-1}$

32.4.3. LEVEL 3

1. Kinetic energy of a particle moving in a straight line varies with time t as K = 4t². The force acting on the particle –
   (A) is constant
   (B) is increasing
   (C) is decreasing
   (D) first increase and then decrease

2. In the figure the block A is released from rest when the spring is at its natural length. For the block B of mass M to leave contact with the ground at some stage, the minimum mass of A must be –
   (A) 2 M
   (B) M
   (C) $\frac{M}{2}$
   (D) a function of M and the force constant of the spring

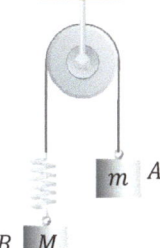

3. The force acting on a body moving along x-axis varies with the position of the particle as shown in the figure. The body is in stable equilibrium at –

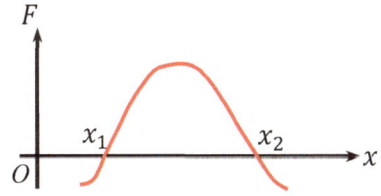

   (A) $x = x_1$ (B) $x = x_2$
   (C) Both x_1 and x_2 (D) Neither x_1 nor x_2

4. A force $\vec{F} = (2\hat{i} + 5\hat{j} + \hat{k})$ is acting on a particle. The particle is first displaced from $(0,0,0)$ to $(2m, 2m, 0)$ along the path $x = y$ and then from $(2m, 2m, 0)$ to $(2m, 2m, 2m)$ along the path $x = 2m$, $y = 2m$. The total work done in the complete path is –

(A) 12 J (B) 8 J (C) 16 J (D) 10 J

MECHANICS

5. A uniform flexible chain of mass m and length $2l$ hangs in equilibrium over a smooth horizontal pin of negligible diameter. One end of the chain is given a small vertical displacement so that the chain slips over the pin. The speed of chain when it leaves pin is-
(A) $\sqrt{2gl}$ (B) $\sqrt{gl}$
(C) $\sqrt{4gl}$ (D) $\sqrt{3gl}$

6. A particle of mass $0.5\ kg$ is displaced from position $\vec{r}_1(2,3,1)$ to $\vec{r}_2(4,3,2)$ by applying of force of magnitude $30\ N$ which is acting along $(\hat{i}+\hat{j}+\hat{k})$. The work done by the force is –
(A) $10\sqrt{3} J$ (B) $30\sqrt{3} J$
(C) 30 J (D) None of these

7. In the given figure, the inclined surface is smooth. The body releases from the top. Then-

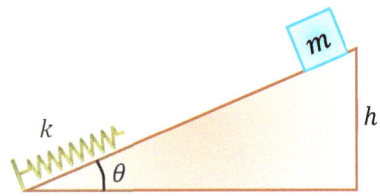

(A) the body has maximum velocity just before striking the spring
(B) The body performs periodic motion
(C) the body has maximum velocity at the compression $\frac{mg \sin \theta}{k}$ where k is spring constant
(D) both (B) and (C) are correct

8. A body of mass $2\ kg$ is moved from a point A to a point B by an external agent in a conservative force field. It the velocity of the body at the points A and B are $5\ m/s$ and $3\ m/s$ respectively and the work done by the external agents is $-10\ J$, then the change in potential energy between points A and B is-
(A) $6\ J$ (B) $36\ J$
(C) $16\ J$ (D) None of these

9. A block of mass M is hanging over a smooth and light pulley through a light string. The other end of the string is pulled by a constant force F. The kinetic energy of the block increases by $20\ J$ in $1s$.
(A) The tension in the string is Mg
(B) The tension in the string is F
(C) The work done by the tension on the block is $20\ J$ in the above $1s$
(D) The work done by the force of gravity is $-20\ J$ in the above $1s$

10. Force acting on a particle is $(2\hat{i}+3\hat{j})\ N$. Work done by this force is zero. when a particle is moved on the line $3y+kx=5$. Here value of k is:
(A) 2 (B) 4 (C) 6 (D) 8

11. A pendulum of mass 1 kg and length $\lambda=1m$ is released from rest at angle $\theta=60º$. The power delivered by all the forces acting on the bob at angle $\theta=30º$ will be: $(g=10\ m/s^2)$
(A) $13.4\ W$ (B) $20.4\ W$
(C) $24.6\ W$ (D) zero

Passage Based Questions –
The system is released from rest with both the springs in unstretched positions. Mass of each block is $1\ kg$ and force constant of each spring is $10\ N/m$.

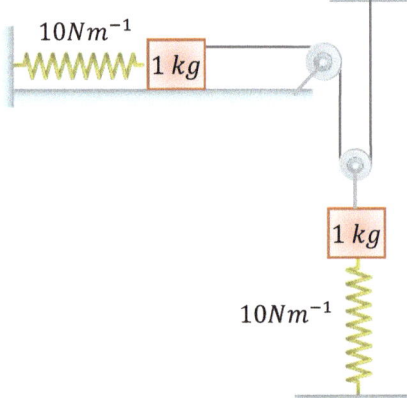

12. Extension of horizontal spring in equilibrium is:
(A) $0.2\ m$ (B) $0.4\ m$ (C) $0.6\ m$ (D) $0.8\ m$

13. Maximum speed for the block placed horizontally is:
(A) $3.21\ m/s$ (B) $2.21\ m/s$
(C) $1.93\ m/s$ (D) $1.26\ m/s$

Assertion & Reason Type Questions –
Each of the questions given below consist of Statement – I and Statement – II. Use the following Key to choose the appropriate answer.
(A) If both Statement- I and Statement- II are true, and Statement - II is the correct explanation of Statement– I.
(B) If both Statement - I and Statement - II are true but Statement - II is not the correct explanation of Statement – I.
(C) If Statement - I is true but Statement - II is false.
(D) If Statement - I is false but Statement - II is true.

14. **Statement - I:** Work done in moving a body in non-uniform circular motion is zero.
    Statement - II: The centripetal force always acts along the radius of the circle.
15. **Statement-I:** Both stretched and compressed springs possess potential energy.
    Statement - II: Work done against restoring force is stored as potential energy.
16. **Statement I:** Work done by or against the friction in moving the body through any round trip is zero.
    Statement II: This is because friction is a conservative force.
17. **Statement I:** Work done in moving a body over a smooth inclined plane does not depend upon slope of inclined plane, provided its height is same.
    Statement II: $W = mgh = mgl\sin\theta$
18. **Statement I:** For the stable equilibrium force has to be zero and potential energy should be minimum.
    Statement II: For the equilibrium it is not necessary that the force is not zero.

    Column Matching Type Questions –
19. Match the following:

Column I	Column II
(A) Work done by all the forces	(P) Change in potential energy
(B) Work done by conservative forces	(Q) Change in kinetic energy
(C) Work done by external forces	(R) Change in mechanical energy
	(S) None

20. A block of mass m is stationary with respect to a rough wedge as shown in figure. Starting from rest in time t, ($m = 1\,kg$, $\theta = 30°$, $a = 2\,m/s^2$, $t = 4s$) work done on block:

    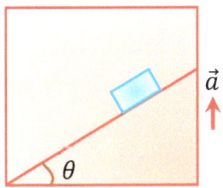

Column-I	Column-II
(A) By gravity	(P) 144 J
(B) By normal reaction	(Q) 32 J
(C) By friction	(R) 56 J
(D) By all the forces	(T) None

32.4.4. LEVEL 4

(Questions asked in previous JEE Mains & Advanced)

SECTION – A [JEE MAIN]

1. A spring of force constant $800\,N/m$ has an extension of $5\,cm$. The work done in extending it from 5 cm to 15 cm is – **[JEE -MAIN 2002]**
   (A) $16\,J$ (B) $8\,J$
   (C) $32\,J$ (D) $24\,J$

2. A spring of spring constant $5\times 10^3\,N/m$ is stretched initially by $5\,cm$ from the unstretched position. Then the work required to stretch it further by another $5\,cm$ is – **[JEE-MAIN 2003]**
   (A) 18.75 N-m (B) 25.00 N-m
   (C) 6.25 N-m (D) 12.50 N-m

3. A uniform chain of length $2\,m$ is kept on a table such that a length of $60\,cm$ hangs freely from the edge of the table. The total of the chain is $4\,kg$. What is the work done in pulling the entire chain on the table– **[JEE-MAIN 2004]**
   (A) 7.2 J (B) 3.6 J
   (C) 120 J (D) 1200 J

4. A force $\vec{F} = 5\hat{i} + 3\hat{j} + 2\hat{k}$ N is applied over a particle which displaces it from its origin to the point $\vec{r} = (2\hat{i} - \hat{j})$m. The work done on the particle in joules is **[JEE-MAIN 2004]**
   (A) -7 (B) $+7$
   (C) $+10$ (D) $+13$

5. A body of mass m is accelerated uniformly from rest to a speed v in a time T. The instantaneous power delivered to the body as a function of time is given by **[JEE-MAIN 2005]**
   (A) $\frac{mv^2}{T^2}.t$ (B) $\frac{mv^2}{T^2}.t^2$
   (C) $\frac{1}{2}\frac{mv^2}{T^2}.t$ (D) $\frac{1}{2}\frac{mv^2}{T^2}.t^2$

6. A mass of M kg is suspended by a weightless string. The horizontal force that is required to displace it until the string makes an angle of $45°$ with the initial vertical direction is – **[JEE MAIN 2006]**
   (A) $\frac{Mg}{\sqrt{2}}$ (B) $Mg(\sqrt{2}-1)$
   (C) $Mg(\sqrt{2}+1)$ (D) $Mg\sqrt{2}$

7. A particle of mass $100\,g$ is thrown vertically upwards with a speed of $5\,m/s$. The work done by

the force of gravity during the time the particle goes up is – **[JEE MAIN 2006]**
(A) 1.25 J   (B) 0.5 J
(C) – 0.5 J   (D) –1.25 J

8. A ball of mass 0.2 kg is thrown vertically upwards by applying a force by hand. If the hand moves 0.2 m while applying the force and the ball goes upto 2 m height further, find the magnitude of the force. Consider $g = 10\ m/s^2$. **[JEE MAIN 2006]**
(A) 20 N   (B) 22 N
(C) 4 N   (D) 16 N

9. The potential energy of a 1 kg particle free to move along the x-axis is given by
$$V(x) = \left(\frac{x^4}{4} - \frac{x^2}{2}\right) \text{ J}.$$
The total mechanical energy of the particle is 2 J. Then, the maximum speed (in m/s) is – **[JEE MAIN 2006]**
(A) $1/\sqrt{2}$ (B) 2
(C) $3/\sqrt{2}$ (D) $\sqrt{2}$

10. A 2 kg block slides on a horizontal floor with a speed of 4 m/s. It strikes a uncompressed spring, and compresses it till the block is motionless. The kinetic friction force is 15 N and spring constant is 10,000 N/m. The spring compresses by- **[JEE MAIN 2007]**
(A) 5.5 cm   (B) 2.5 cm
(C) 11.0 cm (D) 8.5 cm

11. An athlete in the Olympic games covers a distance of 100 m in 10 s. His kinetic energy can be estimated to be in the range **[JEE MAIN 2008]**
(A) $2 \times 10^5 J - 3 \times 10^5 J$
(B) $20,000 J - 50,000 J$
(C) $2,000 J - 5,000 J$
(D) $200 J - 500 J$

12. Consider the following two statements: **[JEE-MAIN 2003]**
(i) Linear momentum of a system of particle is zero.
(ii) Kinetic energy of a system of particle is zero.
Then:
(A) (i) implies (ii) but (ii) does not imply (i)
(B) (i) does not imply (ii) but (ii) implies (i)
(C) (i) implies (ii) and (ii) implies (i)
(D) (i) does not imply (ii) and (ii) does not imply (i)

13. A spherical ball of mass 20 kg is stationary at the top of a hill of height 100 m. It rolls down a smooth surface to the ground, then climbs up another hill of height 30 m and finally rolls down to a horizontal base at a height of 20 m above the ground. The velocity attained by the ball is **[JEE-MAIN 2005]**
(A) 40 m/s   (B) 20 m/s
(C) 10 m/s   (D) $10\sqrt{30}$ m/s

Assertion & Reason Type Questions –
Each of the questions given below consist of Statement – I and Statement – II. Use the following Key to choose the appropriate answer.
(A) If both Statement- I and Statement- II are true, and Statement - II is the correct explanation of Statement– I.
(B) If both Statement - I and Statement - II are true but Statement - II is not the correct explanation of Statement – I.
(C) Statement 1 is true. Statement 2 is false.
(D) Statement 1 is false. Statement 2 is true

14. If two springs S_1 and S_2 of force constants k_1 and k_2 respectively, are stretched by the same force, it is found that more work is done on spring S_1 than S_2.
Statement 1: If stretched by the same amount, work done on S_1 will be more than that on S_2.
Statement 2: $k_1 < k_2$
[JEE MAIN 2012]

15. **Statement-1**: A point particle of mass m moving with speed v collides with stationary point particle of mass M. If the maximum energy loss possible is given as $h\left(\frac{1}{2}mv^2\right)$ then $h = \left(\frac{m}{m+M}\right)$
Statement-2: Maximum energy loss occurs when the particles get stuck together as a result of the collision.
[JEE MAIN 2013]

16. When a rubber-band is stretched by a distance x, it exerts a restoring force of magnitude $F = ax + bx^2$ where a and b are constants. The work done in stretching the unstretched rubber-band by L is **[JEE MAIN 2014]**
(A) $\frac{1}{2}\left(\frac{aL^2}{2} + \frac{bL^3}{3}\right)$ (B) $aL^2 + bL^3$
(C) $\frac{1}{2}(aL^2 + bL^3)$ (D) $\frac{aL^2}{2} + \frac{bL^3}{3}$

17. A person trying to lose weight by burning fat lifts a mass of 10 kg up to a height of 1 m 1000 times.

Assume that the potential energy lost each time he lowers the mass is dissipated. How much fat will he use up considering the work done only when the weight is lifted up? Fat supplies $3.8 \times 10^7 J$ of energy per kg which is converted to mechanical energy with a 20% efficiency rate. Take $g = 9.8\ ms^{-2}$:- **[JEE Main 2016]**

(A) $12.89 \times 10^{-3}\ kg$ (B) $2.45 \times 10^{-3} kg$
(C) $6.45 \times 10^{-3} kg$ (D) $9.89 \times 10^{-3} kg$

18. A point particle of mass m, moves along the uniformly rough track PQR as shown in the figure. The coefficient of friction, between the particle and the rough track equals μ. The particle is released, from rest, from the point P and it comes to rest at a point R. The energies, lost by the particle, over the parts, PQ and PR, of the track, are equal to each other, and no energy is lost when particle changes direction from PQ to QR. The values of the coefficient of friction μ and the distance $x(= QR)$ are, respectively close to:-

[JEE Main 2016]

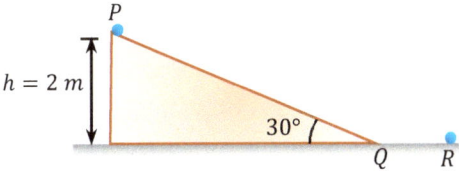

(A) 0.29 and 6.5m (B) 0.2 and 6.5 m
(C) 0.2 and 3.5 m (D) 0.29 and 3.5 m

19. A time dependent force $F = 6t$ acts on a particle of mass $1\ kg$. If the particle starts from rest, the work done by the force during the first 1 sec. will be:
[JEE Main 2017]
(A) $18\ J$ (B) $4.5\ J$
(C) $22\ J$ (D) $9\ J$

20. A particle is moving in a circular path of radius a under the action of an attractive potential $U = -\frac{k}{2r^2}$. Its total energy is **[JEE Main 2018]**
(A) $-\frac{k}{4a^2}$ (B) $\frac{k}{2a^2}$
(C) zero (D) $-\frac{3}{2}\frac{k}{a^2}$

21. A bullet of mass $20g$ has an initial speed of $1 ms^{-1}$ just before it starts penetrating a mud wall of thickness $20\ cm$. If the wall offers a mean resistances of $2.5 \times 10^{-2} N$, the speed of the bullet after emerging from the other side of the wall is close to **[JEE Main 2019]**

(A) $0.7\ ms^{-1}$ (B) $0.3\ ms^{-1}$
(C) $0.1 ms^{-1}$ (D) $0.4 ms^{-1}$

[Although you can solve this problem by application of Newton's laws of motion in presence of friction but here Solve it by applying work energy theorem]

22. A particle which is experiencing a force, given by $\vec{F} = 3\vec{i} - 12\hat{j}$, undergoes a displacement of $\vec{d} = 4\hat{i}$. If the particle had a kinetic energy of $3J$ at the beginning of the displacement, what is its kinetic energy at the end of the displacement?

[JEE Main 2019]

(A) $9 J$ (B) $12 K$
(C) $10 J$ (D) $15 J$

23. A particle moves in one dimension from rest under the influence of a force that varies with the distance traveled by that varies with the distance traveled by the particle as shown in the figure. The kinetic energy of the particle after it has traveled 3 m is:

[JEE Main 2019]

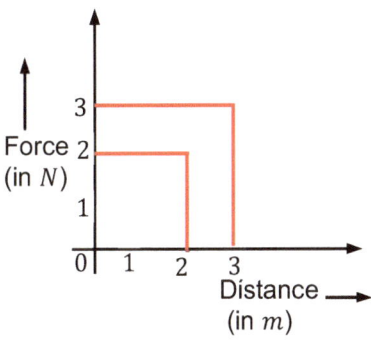

(A) 2.5 J (B) 4 J
(C) 5 J (D) 6.5 J

SECTION - B

1. An elastic string of unstretched length L and force constant k is stretched by a small length x. it is further stretched by another small length y. The work done in the second stretching is – **[IIT - 1994]**
(A) $\frac{1}{2}ky^2$ (B) $\frac{1}{2}k(x^2 + y^2)$
(C) $\frac{1}{2}k(x + y)^2$ (D) $\frac{1}{2}ky(2x + y)$

2. A stone tied to a string of length L is whirled in a vertical circle with the other end of the string at the centre. At a certain instant of time, the stone it at its lowest position, and has a speed u. The magnitude of the change in its velocity as it reaches a position where the string is horizontal is –

[IIT – 1998]

(A) $\sqrt{u^2 - 2gL}$ (B) $\sqrt{2gL}$

(C) $\sqrt{u^2 - gL}$ (D) $\sqrt{2(u^2 - gL)}$

3. A force $\vec{F} = -k(y\hat{i} + x\hat{j})$ (where k is a positive constant) acts on a particle moving in the xy-plane. Starting from the origin, the particle is taken along the positive x-axis to the point $(a, 0)$, and then parallel to the y-axis to the point (a, a). The total work done by the force $\vec{F}$ on the particle is–
   [IIT - 1998]
   (A) $-2ka^2$ (B) $2ka^2$
   (C) $-ka^2$ (D) ka^2

4. A wind-powered generator converts wind energy into electrical energy. Assume that the generator converts a fixed fraction of the wind energy intercepted by its blades into electrical energy. For wind speed v, the electrical power out put will be proportional to –
   [IIT - 2000]
   (A) v (B) v^2
   (C) v^3 (D) v^4

5. A particle, which is constrained to move along the x-axis, is subjected to a force in the same direction which varies with the distance x of the particle from the origin as $F(x) = -kx + ax^3$. Here k and a are positive constants. For $x \geq 0$, the functional form of the potential energy $U(x)$ of the particle is –
   [IIT - 2002]

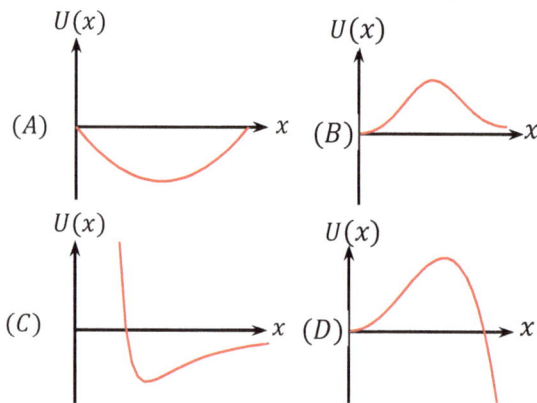

6. An ideal spring with spring-constant k is hung from the ceiling and a block of mass M is attached to its lower end. The mass is released with the spring initially unstretched. Then the maximum extension in the spring is –
   [IIT - 2002]
   (A) $4Mg/k$ (B) $2Mg/k$
   (C) Mg/k (D) $Mg/2k$

7. A block (B) is attached to two unstretched spring S_1 and S_2 with spring constants k and $4k$, respectively (see figure I). The other ends are attached to identical supports M_1 and M_2 not attached to the walls. The springs and supports have negligible masses. There is no friction anywhere. The block B is displaced towards wall 1 by a small distance x (figure II) and released. The block returns and moves a maximum distance y towards wall 2. Displacements x and y are measured with respect to the equilibrium position of the block B. The ratio $\frac{y}{x}$ is –
   [IIT-2008]

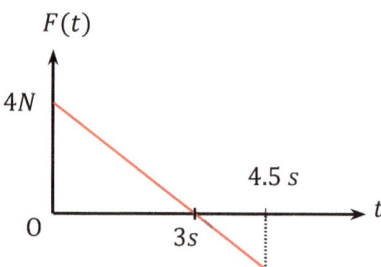

FIGURE

(A) 4 (B) 2 (C) $\frac{1}{2}$ (D) $\frac{1}{4}$

8. A block of mass $2\ kg$ is free to move along the x-axis. It is at rest and from $t = 0$ onwards it is subjected to a time-dependent force $F(t)$ in the x direction. The force $F(t)$ varies with t as shown in the figure. The kinetic energy of the block after 4.5 seconds is
   [IIT 2010]

(A) 4.50 J (B) 7.50 J
(C) 5.06 J (D) 14.06 J

9. The work done on a particle of mass m by a force, $K\left[\frac{x}{(x^2+y^2)^{3/2}}\hat{i} + \frac{y}{(x^2+y^2)^{3/2}}\hat{j}\right]$ (K being a constant of appropriate dimensions), when the particle is taken from the point $(a, 0)$ to the point $(0, a)$ along a circular path of radius a about the origin in the x-y plane is
   [IIT 2013]
   (A) $\frac{2K\pi}{a}$ (B) $\frac{K\pi}{a}$ (C) $\frac{K\pi}{2a}$ (D) 0

10. A wire, which passes through the hole in a small bead, is bent in the form of quarter of a circle. The wire is fixed vertically on ground as shown in the figure. The bead is released from near the top of

the wire and it slides along the wire without friction. As the bead moves from A to B, the force it applies on the wire is [IIT 2014]

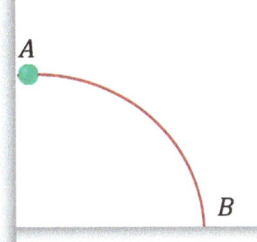

(A) always radially outwards.
(B) always radially inwards.
(C) radially outwards initially and radially inwards later.
(D) radially inwards initially and radially outwards later.

Integer Answer Type Questions

11. A block of mass $0.18\ kg$ is attached to a spring of force-constant $2\ N/m$. The coefficient of friction between the block and the floor is 0.1. Initially the block is at rest and the spring is un-stretched. An impulse is given to the block as shown in the figure. The block slides a distance of 0.06 m and comes to rest for the first time. The initial velocity of the block in m/s is $v = N/10$. Then, N is [IIT 2011]

12. A particle of mass $0.2\ kg$ is moving in one dimension under a force that delivers a constant power $0.5\ W$ to the particle. If the initial speed (in ms^{-1}) of the particle is zero, the speed (in ms^{-1}) after $5\ s$ is [IIT 2013]

13. Consider an elliptically shaped rail PQ in the vertical plane with $OP = 3\ m$ and $OQ = 4\ m$. A block of mass $1\ kg$ is pulled along the rail from P to Q with a force of $18\ N$, which is always parallel to line PQ (see the figure given). Assuming no frictional losses, the kinetic energy of the block when it reaches Q is $(n \times 10)$ joules. The value of n is (take acceleration due to gravity $g = 10\ ms^{-2}$)

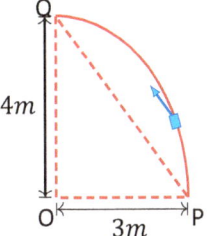

[IIT 2014]

Comprehension Type Questions Passage

14. A particle of mass m is initially at rest at the origin. It is subjected to a force and starts moving along the x-axis. Its kinetic energy K changes with time as $dK/dt = \gamma t$, where γ is a positive constant of appropriate dimensions. Which of the following statements is (are) true? [IIT 2018]
    (A) The force applied on the particle is constant
    (B) The speed of the particle is proportional to time
    (C) The distance of the particle from the origin increases linearly with time
    (D) The force is conservative

15. A ball is projected from the ground at an angle of 45° with the horizontal surface. It reaches a maximum height of $120\ m$ and returns to the ground. Upon hitting the ground for the first time, it loses half of its kinetic energy. Immediately after the bounce, the velocity of the ball makes an angle of 30° with the horizontal surface. The maximum height it reaches after the bounce, in meters, is… [IIT 2018]

16. A particle is moved along a path AB-BC-CD-DE-EF-FA, as shown in figure, in presence of a force $\vec{F} = (ay\hat{\imath} + 2ax\hat{\jmath})N$, Where x and y are in meter and $a = -1\ Nm^{-1}$. The work done on the particle by this force $\vec{F}$ will be ___ Joule. [IIT 2019]

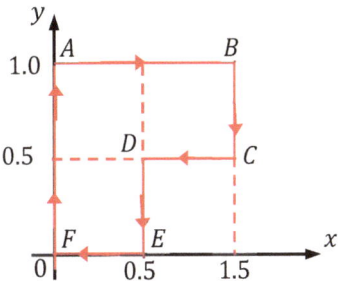

33. ANSWERS KEYS AND SOLUTIONS

33.1. CHECKPOINT 1

1. (a) $62\ kJ$, (b) $-59\ kJ$, (c) $3\ kJ$
2. **APPROACH** The work done by a constant force $\vec{F}$ is defined as
   $W = \vec{F}.\vec{s} = Fs\cos\theta$
   Here, θ is the angle between force $\vec{F}$ and displacement vector $\vec{s}$.
   Now, $0 < \cos\theta < 1$, if $0 < \theta < 90°$,
   $-1 < \cos\theta < 0$, if $90° < \theta < 180°$
   and $\cos\theta = 1$, if $\theta = 0$; $\cos\theta = 0$, if $\theta = 90°$;

$\cos \theta = -1$, if $\theta = 180°$

$\therefore W = Fs \cos \theta = \begin{cases} 0, & \text{if } \theta = \pm 90° \\ +ve, & \text{if } 0 \leq \theta < 90° \\ -ve, & \text{if } 90° < \theta \leq 180° \end{cases}$

SOLUTION

Work done by force $\vec{F}$-

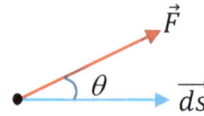

In above diagram, $\vec{F}$ is the force acting on a point and $\vec{ds}$ is its displacement, θ is the angle between force and displacement.

$dW = F\,ds \cos\theta$

$\because$ $0 < \theta < 90°$ $\therefore 0 < \cos\theta < 1$

$\therefore$ $dW = +ve$

i.e., work done by $\vec{F}$ is positive

Work by Friction (f)- Force of friction acts opposite to relative motion

In this case, the angle between force of friction and displacement is $\theta = 180°$ i.e., $\cos\theta = -1$

$\therefore$ $dW = -ve$ i.e., work done by friction $\vec{f}$ will be negative.

WORK BY NORMAL REACTION AND WEIGHT -

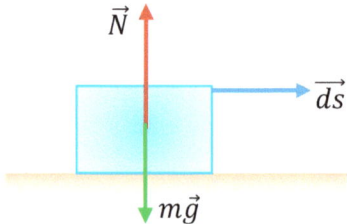

As normal reaction $\vec{N}$ and $m\vec{g}$ both are perpendicular to displacement $\vec{ds}$, therefore the work done by each $\vec{N}$ and $m\vec{g}$ will be zero.

3. Even though the student gets tired (because work is performed within the body to maintain muscles in a state of tension), she does *no work on the book* in merely holding it stationary. She exerts an upward force on the book (equal in magnitude to its weight), but the displacement is zero in this case.

   (b) +44 J

4. The FBD of given problem is shown in following figure-

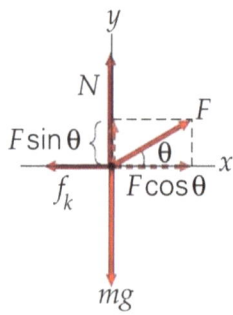

Summing the forces in the x- and y-directions and setting these equal to zero (with a constant velocity $F_{net} = 0$), we get

$\Sigma F_x = F \cos\theta - f_k = F\cos\theta - \mu_k N = ma_x = 0$

$\Sigma F_y = N + F \sin\theta - mg = ma_y = 0$

To find F, the second equation may be solved for N, which is then substituted in the first equation.

$N = mg - F \sin\theta$

(Notice that N is not equal to the weight of the crate. Why?) And, substituting N into the first equation,

$F \cos\theta - \mu_k(mg - F \sin\theta) = 0$

Solving for F and putting the given values:

$F = \dfrac{\mu_k mg}{(\cos\theta + \mu_k \sin\theta)}$

Then,

$W = Fs \cos\theta = \dfrac{\mu_k mg\, s}{(\cos\theta + \mu_k \sin\theta)} \cos\theta$

5. Given: $m = 40.0$ kg

   $\mu_k = 0.550$

   $s = 7.00$ m

   $\theta = 30°$ (from figure)

   v (constant)

   W (work done in moving the crate 7.00 m) =?

   $W = Fs \cos\theta = \dfrac{\mu_k mg\, s}{(\cos\theta + \mu_k \sin\theta)} \cos\theta$

   On substituting the given values, we get-

   $W = 1.15 \times 10^3$ J

6. Work done on the cycle by the road is the work done by the stopping (frictional) force on the cycle due to the road.

   (a) The stopping force and the displacement make an angle of $180°$ (π rad) with each other. Thus, work done by the road,

   $W_r = Fs \cos\theta == 200 \times 10 \times \cos\pi = -2000\,J$

   It is this negative work that brings the cycle to a halt.

   (b) From Newton's Third Law an equal and opposite force acts on the road due to the cycle. Its magnitude is $200\,N$. However, the road undergoes

no displacement. Thus, work done by cycle on the road is zero in a frame attached to ground.

7. Work done $= \vec{F}.\vec{s} = (3\hat{i} + 4\hat{j})N \cdot (3\hat{i} + 4\hat{j})m$
$= (9 + 16)N.m = 25 J$

8. Resolving $\vec{F}_2$ into x and y components-
$F_{2x} = F_2 \cos 60°$
and $F_{2y} = F_2 \sin 60°$

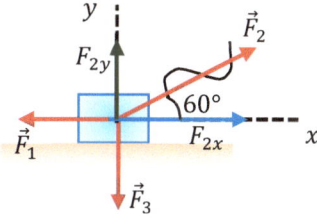

If, $\hat{i}$ and $\hat{j}$ are the unit vectors along x and y directions respectively, then the net force on the trunk
$\vec{F} = \vec{F}_1 + \vec{F}_2 + \vec{F}_3$
$\vec{F} = -F_1\hat{i} + (F_{2x}\hat{i} + F_{2y}\hat{j}) - F_3\hat{j}$
$\vec{F} = -5\hat{i} + 9\cos 60° \hat{i} + 9\sin 60° \hat{j} - 3\hat{j}$
$= -5\hat{i} + \frac{9}{2}\hat{i} + \frac{9\sqrt{3}}{2}\hat{j} - 3\hat{j}$
$= -\frac{1}{2}\hat{i} + \left(\frac{9\sqrt{3}}{2} - 3\right)\hat{j}$
$\vec{s} = -3\hat{i}$

Therefore, net work done on the trunk in ground frame-
$W = \vec{F}.\vec{s} = \left[-\frac{1}{2}\hat{i} + \left(\frac{9\sqrt{3}}{2} - 3\right)\hat{j}\right] \cdot (-3\hat{i})$
$= 1.5 J$

9. (a) Since $\vec{F}$ is in the same direction as the displacement $\vec{s}$, we get:
$W_F = \vec{F} \cdot \vec{s} = Fs \cos 0° = Fs$
The work done by gravity is:
$W_g = m\vec{g} \cdot \vec{s} = mgs \cos(90° + \theta)$
$= -mgs \sin \theta = -mgh$
where $h = y_f - y_i = s \sin \theta$ is the value of the vertical height. That is, the work done by gravity is negative and has a magnitude mg multiplied by height h. This result proves that the work is independent of the path taken between any two points.

Since the force of friction $\vec{f}_k$ is opposite to the displacement $\vec{s}$, $f_k = \mu_k N$, and $N = mg \cos \theta$, the work done by friction will be:
$W_f = \vec{f}_k \cdot \vec{s} = -f_k s = -\mu_k mgs \cos \theta$

Since $\vec{N}$ is perpendicular to $\vec{s}$, we get:

$W_N = \vec{N}.\vec{s} = Ns \cos 90° = 0$

(b) Using the values given, the work done by each force will be:
$W_F = Fs = (20N)(5m) = 100J$
$W_g = -mgs \sin \theta$
$= -(2\text{kg})(9.8 \, ms^{-2})(5\text{m})(\sin 30°)$
$= -49J$
$W_f = -\mu_k mgs \cos \theta$
$= -0.5 \times (2\text{kg})(9.8 \, ms^{-2})(5\text{m})(\cos 30°) = -42.4 \, J$
Thus: $W_\text{net} = W_F + W_g + W_f + W_N$
$= 100 - 49 - 42.4 + 0 = 8.6J$

10. Displacement of the block in $5s$, $s = vt = 2 \times 5 = 10 \, m$.

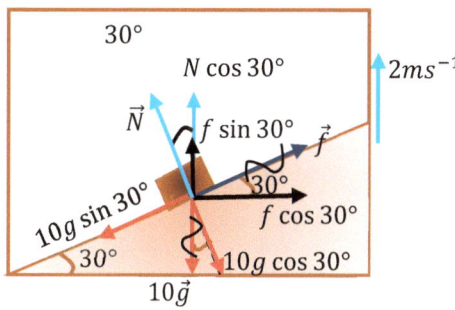

Since, the block is at rest with respect to wedge, therefore, the frictional force on the block,
$f = 10 \, g \sin 30° = 10 \times 10 \times \frac{1}{2} = 50$ newton.
Work done on the block by friction in $5s$,
W_{fr} = component of force of friction in the direction of displacement $\times s$
i.e., $W_{fr} = f \sin 30 \times s = 50 \times \frac{1}{2} \times 10 = 250 \, J$.
Work done on the block by gravitational force in $5s$,
W_{grav} = component of gravitational force in the direction of displacement $\times s$
i.e., $W_{grav} = -10g \times s = -10 \times 10 \times 10 = -1000 \, J$.
Work done on the block by normal reaction in $5s$,
W_N = component of normal reaction in the direction of displacement $\times s$
i.e., $W_N = N \cos 30° \times s$
From above figure, $N = 10g \cos 30° = 10 \times 10 \times \frac{\sqrt{3}}{2}$
$= 50\sqrt{3}$ newton.
$\therefore \quad W_N = 50\sqrt{3} \left(\frac{\sqrt{3}}{2}\right) \times 10 = 750J$

11. $\vec{s} = 3\hat{i} - 2\hat{j} + 4\hat{k}$ and $\vec{F} = 4\hat{i} + 3\hat{j} - \hat{k}$
$W = \vec{F}.\vec{s} = 12 - 6 - 4 = 2$ Joule.

MECHANICS

12. Since, the block is at rest with respect to cart, therefore the acceleration of the block = acceleration of cart = $2ms^{-2}$

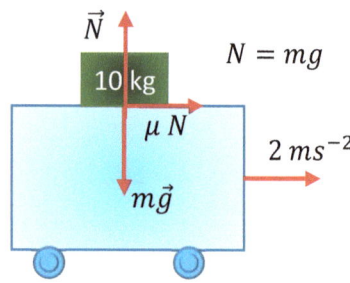

$N = mg$

This acceleration is provided to the block by force of friction between the cart and the block.
The required force of friction corresponding to this acceleration, $f = ma$
i.e., $f = 10 \times 2 = 20N$
The displacement of the block, $s = v_0 t + \frac{1}{2}at^2$
For $t = 2s$, $s = 0 + \frac{1}{2}at^2 = 4m$ ($\because v_0 = 0$)
$W = fs = 20 \times 4 = 80J$

33.2. CHECKPOINT 2

1. **APPROACH** Because F varies with position non-linearly, express the work it does as an integral and evaluate the integral between the limits $x = 1.5\ m$ and $x = 3\ m$:
SOLUTION $W = (c\ N/m^3)\int_{1.5m}^{3m} x^3 dx$
$= (c\ N/m^3)\left[\frac{1}{4}x^{14}\right]_{1.5m}^{3m}$
$= \frac{(cN/m^3)}{4}[(3m)^4 - (1.5m)^4] = 19c\ J$

2. $W = \int_{0.25}^{1.25} e^{-4x^2} dx = 0.21\ J$

3. **APPROACH** We can express the mass of the water in the bucket as the difference between its initial mass and the product of the rate at which it loses water and her position during her climb. Because the person must do work against gravity in lifting and carrying the bucket, the work he does is the integral of the of the weight of the bucket as a function of its position.
SOLUTION(a) Express the mass of the bucket and the water in it as a function of its initial mass, the rate at which it is losing water, and the person's position, y, during her climb:
$m(y) = 40\text{kg} - ry$

Find the rate, $r = \frac{\Delta m}{\Delta y}$, at which the person's bucket loses water: $r = \frac{\Delta m}{\Delta y} = \frac{20\text{kg}}{20\text{m}} = 1\text{kg/m}$

$\therefore\ m(y) = 40\text{kg} - ry = \sqrt{40\text{kg} - \frac{1\text{kg}}{m}y}$

(b) Integrate the force person exerts on the bucket, $m(y)g$, between the limits of $y = 0$ and $y = 20\ m$:
$W = g\int_0^{20}(40\text{kg} - \frac{1kg}{m}y)\,dy$
$= (9.81\text{m/s}^2)[(40\text{kg})y - \frac{1}{2}(1\text{kg/m})y^2]_0^{20}$
$= 5.89\ kJ$

33.3. CHECKPOINT 3

1. The normal force balances the weight of the block, and neither of these forces does work since the displacement is horizontal. Since there is no friction, the resultant external force is the $12\ N$ force. The work done by this force is

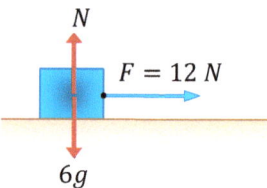

$W = Fs = (12N)(3.0m) = 36\ Nm = 36\ J$
Using the work-energy theorem and noting that the initial kinetic energy is zero, we get
$W = K_f - K_1 = \frac{1}{2}mv_f^2 - 0$
$v_f^2 = \frac{2W}{m} = \frac{2(36J)}{6.0\ kg} = 12m^2/s^2$
$v_f = 3.5\text{m/s}$

2. The forces acting on the ball are-
(1) Gravitational force $m\vec{g}$ in downward direction,
(2) Air resistance opposite to velocity of the ball.
Work done by gravitational force on the ball
$W_{grav} = -mg(h_2 - h_1)$
Work done on the ball by air resistance is $W_{air} = ?$
From work kinetic energy theorem, we have-
$W_{total} = K_2 - K_1$
Here, $W_{total} == W_{grav} + W_{air}$
$\therefore W_g + W_{air} = K_2 - K_1$
$\Rightarrow -mg(h_2 - h_1) + W_{air} = \frac{1}{2}m(v_2^2 - v_1^2)$
$\Rightarrow W_{ar} = mg(h_2 - h_1) + \frac{1}{2}m(v_2^2 - v_1^2)$

3. By work energy theorem, we have-
$W_{total} = K_2 - K_1$
Here, the forces acting on the particle are-

(1) Gravitational force F_{grav} in downward direction,
(2) Air resistance F_{air} opposite to velocity of the body.

$\therefore W_{total} = W_{grav} + W_{air} = K_2 - K_1$

$-mgh + W_{air} = 0 - \frac{1}{2}mv_0^2$

or $W_{air} = mgh - \frac{1}{2}mv_0^2 = m[gh - \frac{1}{2}v_0^2]$

or $W_{air} = (1\ kg)[(10\ m/s^2)(4m) - \frac{1}{2}(10\ m/s)^2]$

or $W_{air} = (1\ kg)[40\ m^2/s^2 - 50\ m^2/s^2]$

or $W_{air} = -10\ J$

4. As in projectile motion, the horizontal component of velocity does not change, therefore

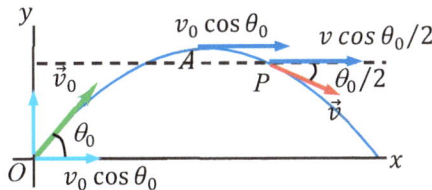

$v_0 \cos\theta_0 = v \cos\theta_0/2$

$v = \frac{v_0 \cos\theta_0}{\cos\theta_0/2}$ ... (1)

The net force acting on the particle is only the gravitational force, therefore, from work energy theorem, we have

$W_{total} = W_{grav} = K_2 - K_1$ ... (2)

Here, K_1 = KE of the particle at highest point A,

$K_1 = \frac{1}{2}m(v_0 \cos\theta_0)^2 = \frac{1}{2}mv_0^2 \cos^2\theta_0$ ... (3)

K_2 = KE of the particle at point P,

$K_2 = \frac{1}{2}mv^2 = \frac{1}{2}m\left(\frac{v_0 \cos\theta_0}{\cos\theta_0/2}\right)^2$ ... (4)

Substituting these values of K_1 and K_2 in (2), we get

$W_{grav} = K_2 - K_1$

$= \left[\frac{1}{2}m\left(\frac{v_0 \cos\theta_0}{\cos\theta_0/2}\right)^2 - \frac{1}{2}mv_0^2 \cos^2\theta_0\right]$

$= \frac{1}{2}mv_0^2 \cos^2\theta_0 \left[\left(\frac{1}{\cos\theta_0/2}\right)^2 - 1\right]$

$= \frac{1}{2}mv_0^2 \cos^2\theta_0 \left[\frac{1-\cos^2\theta_0/2}{\cos^2\theta_0/2}\right]$

$= \frac{1}{2}mv_0^2 \cos^2\theta_0 \left[\frac{\sin^2\theta_0/2}{\cos^2\theta_0/2}\right]$

$= \frac{1}{2}mv_0^2 \cos^2\theta_0 \tan^2\theta_0/2$

5. $\mu_k = 0.35$

6. **APPROACH** Solve the problem by using the work-kinetic energy theorem. According to it-

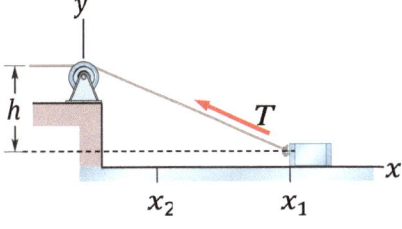

$W_{total} = K_2 - K_1$

SOLUTION The force acting on the cart, are-
(i) Tension in the cart T,
(ii) Downward gravitational force, and
(iii) Normal reaction of the surface

As gravitational force, and normal reaction are perpendicular to the displacement, therefore the work done by these two forces will be zero. Therefore,

$W_{total} = W_T = K_2 - K_1$ ... (1)

Here, $W_T = T.d$, where $T = 25\ N$, is the tension in the cord and d is the length of the cord pulled as the cart slides from x_1 to x_2. From the figure, we have

$d = \sqrt{x_1^2 + h^2} - \sqrt{x_2^2 + h^2}$

$= \sqrt{(3.00m)^2 + (1.20m)^2} - \sqrt{(1.00m)^2 + (1.20m)^2}$

$= 3.23m - 1.56m = 1.67m$

$\therefore W_T = T.d = (25\ N)(1.67m) = 41.75J$

Substituting, this value in (1), we get-

$41.75J = K_2 - K_1$

Therefore, the change in KE of the cart-

$\Delta K = 41.75J$

7. Both of these claims are false. Question (a) refers to a simple, everyday experience that unfortunately cannot be analysed by means of traditional physics teaching without the introduction of additional work-like quantities and energy-like equations. The upward force on the boy that projects him into the air is the normal force on his feet from the ground. The centre of mass of the boy indeed moves through an upward displacement. The normal force, however, goes through no displacement in the reference frame of the ground, and therefore no work is done by this force on the boy. The change in the boy's kinetic energy does not come from work done on the system of the boy. This is a case of a deformable system. Other cases include a person climbing stairs or a ladder, a girl pushing off a wall while standing on a skateboard, and a piece of putty slamming into a wall. In all of these cases, no work is done by the contact force, because there is no displacement of the point of application of the force.

33.4. CHECKPOINT 3

1. (a) APPROACH Once we have chosen the reference point for $y = 0$, we can calculate the gravitational potential energy U of the system *relative to that reference point* with $U = mgy$.

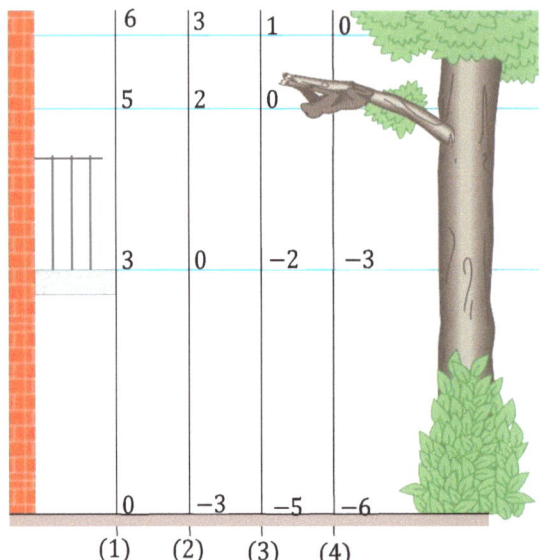

SOLUTION For choice (1) the sloth is at $y = 5.0\, m$, and $U = mgy = (2.0\text{kg})(9.8\, m/s^2)(5.0\text{m}) = 98\, J$
For the other choices, the values of U are
$U = mgy = mg(2.0\text{m}) = 39\text{J}$
$U = mgy = mg(0) = 0\text{J}$
$U = mgy = mg(-1.0\text{m})$
$\quad = -19.6\text{J} \approx -20\text{J}$

(b) APPROACH The *change* in potential energy does not depend on the choice of the reference point for $y = 0$; instead, it depends on the change in height Δy.

SOLUTION For all four situations, we have the same $\Delta y = -5.0\, m$. Thus, for (1) to (4), the equation $U = mgy$ gives that-
$\Delta U = mg\Delta y = (2.0\text{kg})(9.8\, m/s^2)(-5.0\text{m}) = -98\, J$

2. (a) The system must have at least two objects that interact with each other through one of the conservative forces we know, gravity or the spring force: The system includes the hammer and the Earth. The force between them is gravity. (b) $-47\, J$ (c) $-47\, J$

3. (a) System: ball, spring (and the ceiling) and the earth.
(b) $\Delta U = -mgh + \frac{1}{2}kh^2 + khy_i$

4. (a) The gravitational potential energy increases until it reaches its maximum value when the stone reaches its highest point above the ground. (b) The kinetic energy decreases as the potential energy increases. It is zero at the highest point. (c) The force of gravity does work on the stone throughout its motion.

33.5. CHECKPOINT 4

1. $0.0010\, J$

2. APPROACH The only forces doing work on the block are the force of gravity and the elastic restoring force of the spring, both of which are conservative forces. Since there are no non-conservative forces, therefore, we can apply the principle of conservation of mechanical energy:

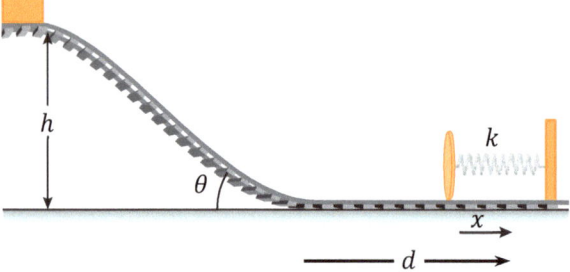

SOLUTION According to principle of conservation of mechanical energy, we have-
$$U_i + K_i = U_f + K_f \quad \ldots (1)$$
The block starts at rest, and when the spring is at maximum compression ($= x$, say), the block is also momentarily at rest. Therefore, the initial and the final kinetic energies of the block and spring system are zero, i.e., $K_i = K_f = 0$.
Considering the reference potential energy level at ground, we have
$$U_i = (U_{grav})_i + (U_{spring})_i = mgh + 0$$
$$U_f = (U_{grav})_f + (U_{spring})_f = 0 + \frac{1}{2}kx^2$$
Substituting these values in (1), we get-
$$mgh + 0 = \left(0 + \frac{1}{2}kx^2\right) + 0$$
Solving for x and substituting, we get
$$x = \sqrt{\frac{2mgh}{k}} = \sqrt{\frac{2(1.2\text{kg})(9.81\text{m/s}^2)(2.3\text{m})}{460\text{N/m}}} = 0.34\text{m}$$

3. (a) The mass of the element $dm = \left(\frac{m}{l}\right)R d\theta$
The gravitational PE of the element $dU = (dm)gy$
Thus, the gravitational potential energy of whole chain
$$U = \int (dm)gy = \int_0^{l/R} \left(\frac{m}{l}R d\theta\right) g(R\cos\theta)$$
$$= \frac{mR^2 g}{l} \int_0^{l/R} \cos\theta\, d\theta = \frac{mgR^2}{l}[\sin\theta]_0^{l/R}$$
$$= \frac{mg}{2}^2 \sin\left(\frac{l}{R}\right)$$

WORK, ENERGY AND POWER   81

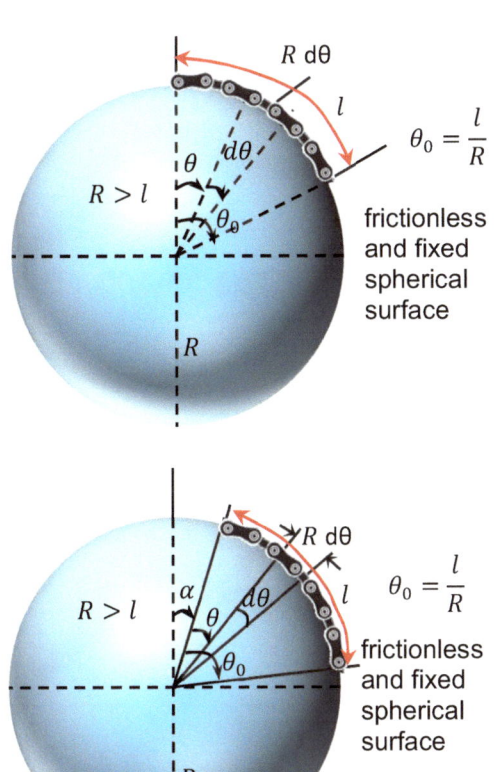

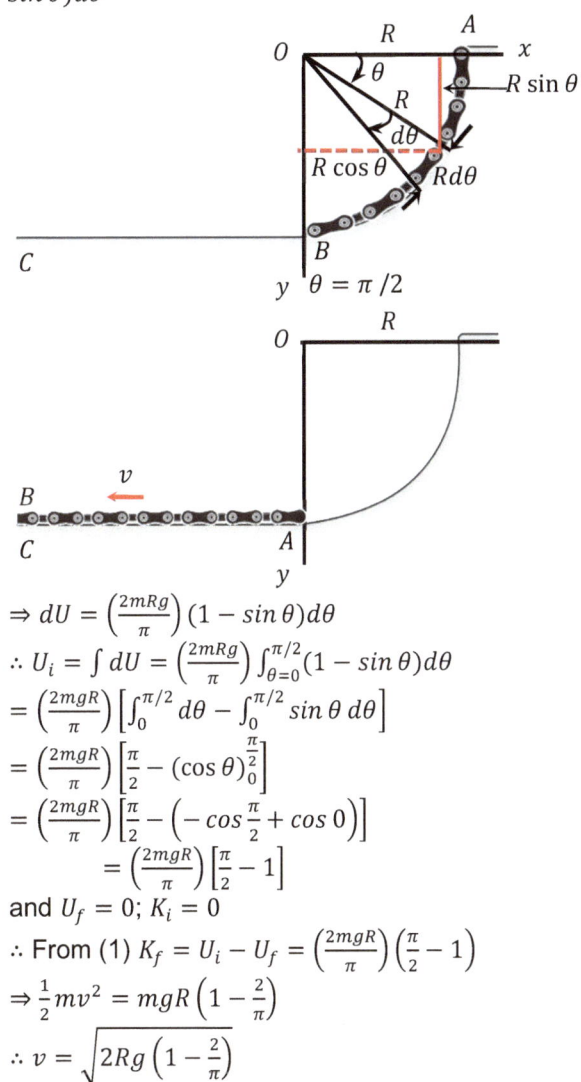

$\Rightarrow U_i - U_f = K_f - K_i = \frac{1}{2}mv^2 - 0$ ... (1)

$dU = (dm)gh = (\lambda R d\theta)g(R - R\sin\theta) = \frac{mR^2g}{\pi R/2}(1 - \sin\theta)d\theta$

(b) (Total mechanical energy)$_i$
= (Total mechanical energy)$_f$

$\Rightarrow U_i + K_i = U_f + K_f$

$\Rightarrow K_f = U_i - U_f$

$\therefore U_i = \int dU = \frac{mR^2g}{l}\int_0^{l/R}\cos\theta\, d\theta$

$= \frac{mg}{l}R^2[\sin\theta]_0^{l/R} = \frac{mgR^2}{l}\sin\left(\frac{l}{R}\right)$

and $U_f = \int dU = \frac{mR^2g}{l}\int_\alpha^{\alpha+\theta_0}\cos\theta\, d\theta$

$= \frac{mg}{l}R^2[\sin\theta]_{\theta=\alpha}^{\alpha+\theta_0}$

$= \frac{mgR^2}{l}\left[\sin\left(\alpha+\frac{l}{R}\right) - \sin\alpha\right]$

$\therefore K_f = U_i - U_f = mg\frac{R^2}{l}\left[\sin\left(\frac{l}{R}\right) - \sin\left(\alpha+\frac{l}{R}\right) + \sin\alpha\right]$

$\Rightarrow \frac{1}{2}mv_f^2 = mg\frac{R^2}{l}\left[\sin\left(\frac{l}{R}\right) - \sin\left(\alpha+\frac{l}{R}\right) + \sin\alpha\right]$

$\Rightarrow v_f = R\sqrt{\frac{2g}{l}\left[\sin\left(\frac{l}{R}\right) - \sin\left(\alpha+\frac{l}{R}\right) + \sin\alpha\right]}$

(c) Tangential force on dm

$= (dm)g\sin\theta = \left(\frac{mRg}{l}\right)\sin\theta\, d\theta$

Tangential force on the chain

$= \left(\frac{mRg}{l}\right)\int_{\theta=0}^{\frac{l}{R}}\sin\theta\, d\theta = \left(\frac{mRg}{l}\right)[-\cos\theta]_0^{\frac{l}{R}}$

$= \left(\frac{mRg}{l}\right)\left[1 - \cos\frac{l}{R}\right]$

Tangential acceleration force on the chain

$= \left(\frac{Rg}{l}\right)\left[1 - \cos\frac{l}{R}\right]$

4. $(E_{mech})_i = (E_{mech})_f$

$\Rightarrow KE_i + PE_i = KE_f + PE_f$

$\Rightarrow dU = \left(\frac{2mRg}{\pi}\right)(1 - \sin\theta)d\theta$

$\therefore U_i = \int dU = \left(\frac{2mRg}{\pi}\right)\int_{\theta=0}^{\pi/2}(1-\sin\theta)d\theta$

$= \left(\frac{2mgR}{\pi}\right)\left[\int_0^{\pi/2}d\theta - \int_0^{\pi/2}\sin\theta\, d\theta\right]$

$= \left(\frac{2mgR}{\pi}\right)\left[\frac{\pi}{2} - (\cos\theta)_0^{\frac{\pi}{2}}\right]$

$= \left(\frac{2mgR}{\pi}\right)\left[\frac{\pi}{2} - \left(-\cos\frac{\pi}{2} + \cos 0\right)\right]$

$= \left(\frac{2mgR}{\pi}\right)\left[\frac{\pi}{2} - 1\right]$

and $U_f = 0$; $K_i = 0$

$\therefore$ From (1) $K_f = U_i - U_f = \left(\frac{2mgR}{\pi}\right)\left(\frac{\pi}{2} - 1\right)$

$\Rightarrow \frac{1}{2}mv^2 = mgR\left(1 - \frac{2}{\pi}\right)$

$\therefore v = \sqrt{2Rg\left(1 - \frac{2}{\pi}\right)}$

5. Initial kinetic energy, $KE_i = 0$;

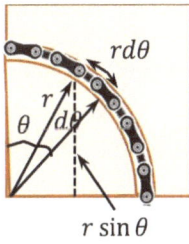

Calculation of initial potential energy:

$dU_i = (\lambda r d\theta)gh$

$\Rightarrow dU_i = \frac{m}{(\pi r/2)}(rd\theta)g(r\cos\theta) = \left(\frac{2mgr}{\pi}\right)\cos\theta\, d\theta$

$$\therefore \quad U_i = \int dU =$$
$$\left(\tfrac{2mgr}{\pi}\right)\int_{\theta=0}^{\pi/2}\cos\theta\, d\theta =$$
$$\left(\tfrac{2mgr}{\pi}\right)[\sin\theta]_0^{\pi/2} = \left(\tfrac{2mgr}{\pi}\right)$$

Calculation of final potential energy:
$$dU = (\lambda dx)g(-x) = \left(\tfrac{-mg}{\pi r/2}\right)x\, dx$$

$$\therefore\quad U_f = \int dU = -\left(\tfrac{2mg}{\pi r}\right)\int_{x=0}^{\pi r/2}x\, dx = -\left(\tfrac{2m}{\pi r}\right)\left(\tfrac{\pi r}{2}\right)^2\tfrac{1}{2} =$$
$$-\left(\tfrac{2mg}{\pi r}\right)\tfrac{(\pi r)^2}{8} = \tfrac{-mg(\pi r)}{4}$$

$\therefore$ By conservation of mechanical energy
$$KE_i + U_i = KE_f + U_f$$
$$\Rightarrow KE_f = U_i - U_f = \left(\tfrac{2mgr}{\pi}\right) - \tfrac{-mg(\pi r)}{4} = (mgr)\left(\tfrac{2}{\pi}+\tfrac{\pi}{4}\right)$$
$$\therefore \tfrac{1}{2}mv^2 = (mgr)\left(\tfrac{2}{\pi}+\tfrac{\pi}{4}\right)$$
$$\therefore v = \sqrt{2gr\left(\tfrac{2}{\pi}+\tfrac{\pi}{4}\right)}$$

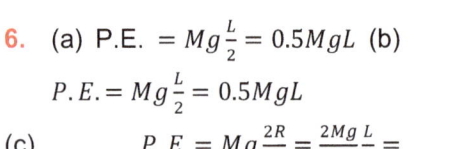

6. (a) P.E. $= Mg\tfrac{L}{2} = 0.5MgL$ (b)
   $P.E. = Mg\tfrac{L}{2} = 0.5MgL$

   (c) $\quad P.E. = Mg\tfrac{2R}{\pi} = \tfrac{2Mg}{\pi}\tfrac{L}{\pi} = \tfrac{2}{\pi^2}MgL \approx 0.2MgL$

   (d) $P.E. = \tfrac{4}{\pi^2}MgL \approx 0.4MgL$

   (e) $P.E. = Mg\left(R-\tfrac{2R}{\pi}\right) = \left(\tfrac{\pi-2}{\pi}\right)MgR$
   $= \tfrac{\pi-2}{\pi^2}MgL = \tfrac{1.14}{\pi^2}MgL = 0.11MgL$

   It is clear that: (e) < (c) < (d) < (a) = (b)

7. For given PE vs x graph, the force vs x graph shown in Fig.1b. Points x_1 and x_3 are stable equilibrium points. At both points, $F_x = -\tfrac{dU}{dx}$ is zero because the slope of the $U(x)$ curve is zero. When the particle is displaced to either side, the force pushes back toward the equilibrium point. The slope of the $U(x)$ curve is also zero at points x_2 and x_4, and these are also equilibrium points. But when the particle is displaced a little to the right of either point, the slope of the $U(x)$ curve becomes negative, corresponding to a positive F_x that tends to push the particle still farther from the point. When the particle is displaced a little to the left, F_x is negative, again pushing away from equilibrium. This is analogous to a marble rolling on the top of a bowling ball. Points x_2 and x_4 are called **unstable** equilibrium points; *any maximum in a potential-energy curve is an unstable equilibrium position.*

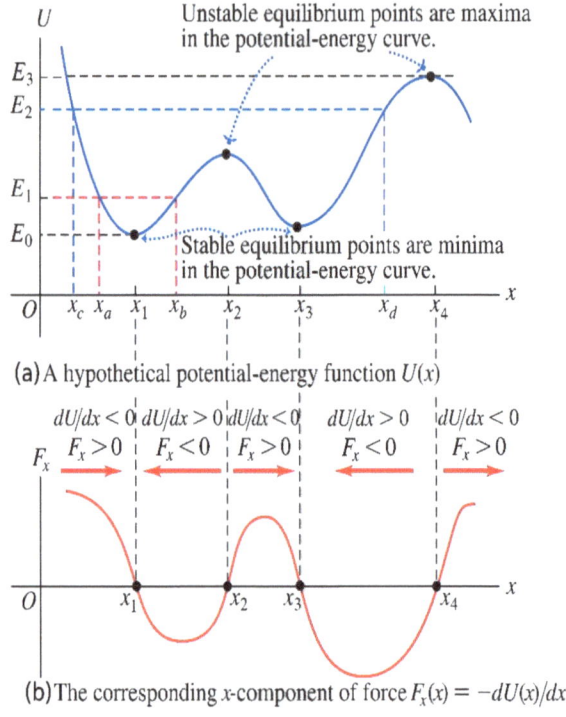

(a) A hypothetical potential-energy function $U(x)$

(b) The corresponding x-component of force $F_x(x) = -dU(x)/dx$

FIGURE 1

If the total energy is E_1 and the particle is initially near x_1, it can move only in the region between x_a and x_b determined by the intersection of the E_1 and U graphs (Fig. a). Again, U cannot be greater than E_1 because K can't be negative. We speak of the particle as moving in a *potential well,* and x_a and x_b are the *turning points* of the particle's motion (since at these points, the particle stops and reverses direction). If we increase the total energy to the level E_2, the particle can move over a wider range, from x_c to x_d. If the total energy is greater than E_3, the particle can "escape" and move to indefinitely large values of x. At the other extreme, E_0 represents the minimum total energy the system can have.

☞ If the total energy $E > E_3$, the particle can "escape" to $x > x_4$.
If $E = E_2$, the particle is trapped between x_c and x_d
If $E = E_1$, the particle is trapped between x_a and x_b
Minimum possible energy is E_0 the particle is at rest at x_1.

8. The force applied by the motor, through the cable, is the tension force $\vec{T}$. This force does work on the machine with power $P = Tv$. The machine is in equilibrium, because the motion is at constant

velocity, hence the tension in the rope balances the friction and is
$T = f_k = \mu_k mg$
The motor's power output is
$P = Tv = \mu_k mgv = 882W$

9. The net force on a car moving at a steady speed is zero. The motion is opposed both by rolling friction and by air resistance. The forward force on the car $\vec{F}_{car}$ (actually it is the force applied by ground on car, a reaction to the drive wheels pushing backward on the ground) exactly balances the two opposing forces:
$F_{car} = f_r + F_{drag}$
$F_{car} = \mu_r mg + F_{drag} = 294N + 655N = 949N$
$P_{car} = F_{car}v = (949N)(30\,m/s) = 28,500W = 38\,hp$
This is the power needed at the drive wheels to push the car against the dissipative forces of friction and air resistance. The power output of the engine is larger because some energy is used to run the water pump, the power steering, and other accessories. In addition, energy is lost to friction in the drive train. If 25% of the power is lost (a typical value), leading to $P_{car} = 0.75 P_{engine}$, the engine's power output is
$P_{engine} = \frac{P_{car}}{0.75} = 38,000W = 51hp$

33.6. CONCEPTUAL QUESTIONS

1. No. Work requires that a force acts through a distance.
2. False. Any force acting on an object can do work. The work done by different forces may add to produce a greater net work, or they may cancel to some extent. It follows that the net work done on an object can be thought of in the following two equivalent ways: (i) The sum of the work done by each individual force; or (ii) the work done by the net force.
3. True. To do work on an object a force must have a nonzero component along its direction of motion.
4. There is not enough information to tell. The lost potential energy and the work done by the environment could increase the kinetic energy or it is possible that all the work and energy are converted to thermal energy
5. There is not enough information to tell. The work done could cause some or all of the potential energy change or some of the work could be converted to thermal energy. Without more information, it is impossible to say whether a kinetic energy change is present
6. The system is doing work on the environment. The total mechanical energy of the system is lower.
7. The ball's kinetic energy is equal to the work done on it by gravity. Since work is force × displacement, the kinetic energy of the ball increases by equal amounts in equal distance intervals
8. Yes, she is doing work. The work done by her and the work done on her by the river are opposite in sign, so they cancel and she does not move with respect to the shore. When she stops swimming, the river continues to do work on her, so she floats downstream.
9. No work was done by gravity. $W_g = -m_g \Delta y$. Here, $\Delta y = 0$. Any work done during a downward part of the motion was undone during the upward parts.
10. No, not if the object is moving in a circle. Work is the product of force and the displacement in the direction of the force. Therefore, a centripetal force, which is perpendicular to the direction of motion, cannot do work on an object moving in a circle.
11. You are doing no work on the wall. Your muscles are using energy generated by the cells in your body and producing byproducts which make you feel fatigued.
12. Yes. The magnitudes of the force, displacement and the angle between them are the relevant quantities, and displacement depends on the choice of coordinate system.
13. Yes. When you are in an elevator, which is moving with respect to ground, then work done by normal reaction applied on you with respect to ground is positive when it is moving upwards and negative when it is moving in downward direction.
14. The bullet with the smaller mass has a speed which is greater by a factor of $\sqrt{2} = 1.4$. Since their kinetic energies are equal, then $\frac{1}{2}m_1 v_1^2 = \frac{1}{2}m_2 v_2^2$ If $m_2 = 2m_1$, then $\frac{1}{2}m_1 v_1^2 = \frac{1}{2}(2m_1)v_2^2$, so $v_1 = \sqrt{2}v_2$. They can both do the same amount of work, however, since their kinetic energies are the same.

15. The net work done on a particle and the change in the kinetic energy are independent of the choice of reference frames only if the reference frames are at rest with respect to each other. The work-energy principle is also independent of the choice of reference frames if the frames are at rest with respect to each other.

    If the reference frames are in relative motion, the net work done on a particle, the kinetic energy, and the change in the kinetic energy all will be different in different frames. The work-energy theorem will still be true.

16. The two forces on the book are the applied force upward (nonconservative) and the downward force of gravity (conservative). If air resistance is non-negligible, it is nonconservative.

17. (a) If the net force is conservative, the change in the potential energy is equal to the negative of the change in the kinetic energy, so $\Delta U = -300\ J$. (b) If the force is conservative, the total mechanical energy is conserved, so $\Delta E = 0$.

18. The kinetic energy cannot be negative, since m and v^2 are always positive or zero. The gravitational potential energy can be negative since any level can be chosen to be zero.

19. No. The maximum height on the rebound cannot be greater than the initial height if the ball is dropped. Initially, the dropped ball's total energy is gravitational potential energy. This energy is changed to other forms (kinetic as it drops, and elastic potential during the collision with the floor) and eventually back into gravitational potential energy as the ball rises back up. The final energy cannot be greater than the initial (unless there is an outside energy source) so the final height cannot be greater than the initial height. Note that if you *throw* the ball down, it initially has kinetic energy as well as potential so it may rebound to a greater height.

20. Yes. Suppose two blocks A and B are placed on a smooth surface as shown in following figure. Let the contact surfaces between A and B are rough.

    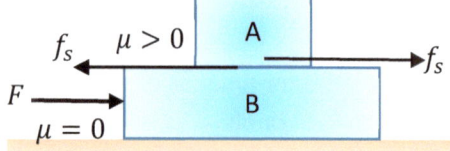

    Now, suppose we push lower block B in forward direction such that the upper block A also moves in forward direction without any relative slipping between them, then there will be a static friction between A and B and the direction of the force of static friction will be in forward direction on upper block A, and in backward direction on lower block B. In this case, the work done by static friction on the upper block A, in a frame attached to ground is positive. As in this case the upper block is at rest with respect to lower block B, therefore the work done by static friction on A, as seen from the frame of lower block B is zero.

    As the force of static friction on the lower block is in backward direction, therefore in ground's frame the work done by static friction is negative. Since, there is no relative slipping between A and B, therefore from the frame attached to A, the work done by force of friction on A is zero.

    Now, suppose there is a relative slipping between blocks A and B. In this case, the force of friction between A and B will be kinetic in nature. The upper block A slips in backward direction with respect to the lower block B, whereas lower block B slips in forward with respect to upper block A. Therefore, the work done by kinetic friction on the upper block A, in a frame attached to ground, will be positive (because in ground's frame, the upper block is moving in forward direction).

    Work done by force of kinetic friction on lower block A, in a frame attached to ground will be negative.

    Work done by force of kinetic friction on upper block A in a frame attached to lower block B will be negative.

    And the work done by force of kinetic friction on lower block B, in a frame attached to upper block A, will be positive

21. No work is done *on the wall* (since the wall does not undergo displacement) but internally your muscles are converting chemical energy to other forms of energy, which makes you tired.

22. Yes, the work done by the net force acting on a particle depends on the inertial reference frame of the observer.

    Yes, the change in kinetic energy so depend. Newton's laws are valid only in inertial frames of reference. If we find Newton's second law to hold in one frame of reference, then it holds in all inertial frames. If two observers in different inertial frames move at constant velocity v relative to one another and observe the same experiment, they measure identical values for the forces, masses, and accelerations, and so they agree completely in their analysis using Newton's second law. In the Newton's second law, observers in different inertial frames will agree on the results of applying the work energy theorem. However, unlike forces and accelerations, displacements and velocities measured by observers in different inertial frames will, in general, be different, and so they will deduce different values for the work and kinetic energies in the experiment. Therefore, both the value of work and kinetic energy depends on the reference frame of the observer. Let us consider an observer is at Mars observes the movement of two object which is present at Earth's surface. Both the observers observe to each other at rest in the Earth's surface. But according to the observer which is at Mars observes both the objects are moving. Therefore, the value of kinetic energy depends on the reference frame of the observer. Again, since the net work done by the forces acting on a body is equal to the change in the kinetic energy of the body, therefore the value of work depends on the reference frame of the observer.

23. Yes, the spring can leave the table. When you push down on the spring, you do work on it and it gains elastic potential energy, and loses a little gravitational potential energy, since the center of mass of the spring is lowered. When you remove your hand, the spring expands, and the elastic potential energy is converted into kinetic energy and into gravitational potential energy. If enough elastic potential energy was stored, the center of mass of the spring will rise above its original position, and the spring will leave the table.

24. Yes, For, example: rocket flying in space gains or loses speed by ejecting high speed gases.

25. The initial potential energy of the water is converted first into the kinetic energy of the water as it falls. When the falling water hits the pool, it does work on the water already in the pool, creating splashes and waves. Additionally, some energy is converted into heat and sound.

26. The elastic potential energy of the compressed spring is converted into thermal energy, which is dispersed within the solution.

27. There is no violation. Suppose you pick a book of mass m from floor and put it on a table of height h, then work done by you will be $W_{you} = +mgh$, whereas the work done by gravity will be $W_{grav} = -mgh$. Therefore, net work done on the book by all the forces $W_{tot} = +mgh - mgh = 0$.
∴ from work energy theorem, $W_{tot} = K_2 - K_1$
$\Rightarrow 0 = K_2 - K_1$ or $K_2 = K_1$

28. (a) As a car accelerates uniformly from rest, the potential energy stored in the fuel is converted into kinetic energy in the engine and transmitted through the transmission into the turning of the wheels, which causes the car to accelerate (if friction is present between the road and the tires).
(b) If there is a friction force present between the road and the tires, then when the wheels turn, the car moves forward and gains kinetic energy. If the static friction force is large enough, then the point of contact between the tire and the road is instantaneously at rest – it serves as an instantaneous axis of rotation. If the static friction force is not large enough, the tire will begin to slip, or skid, and the wheel will turn without the car moving forward as fast. If the static friction force is very small, the wheel may spin without moving the car forward at all, and the car will not gain any kinetic energy (except the kinetic energy of the spinning tires).

29. The work-kinetic energy theorem is only applicable to a system which can be treated like a particle. We cannot treat a sliding block acted on by a frictional force as a particle (even though we *can* continue to treat it as a particle, as we did, when analysing its behaviour using Newton's laws) The frictional force, which we represented as a constant force is in reality quite complicated, involving the making and breaking of many microscopic welds which deform the surfaces and result in changes in internal energy of the surfaces (which may in part be revealed as an increase in the temperature of the surfaces). Because of the difficulty of accounting for these other forms of energy, and because the objects do not behave as particles, it is generally not correct to apply the particle form of

the work–energy theorem to objects subject to frictional forces.

Note: Generally, to solve numerical problems, we consider a sliding body over a frictional surface as a particle.

30. No. A body cannot have kinetic energy without having momentum.
The relation between kinetic energy K and linear momentum (p) is, $K = \frac{p^2}{2m}$
Here, p is a vector quantity. if, $p = 0$, then $K = 0$.
Therefore, it is not possible a body has kinetic energy without having momentum.
(b) Yes. A body can have momentum without having kinetic energy.
The momentum (p) of photon which has zero rest mass of electromagnetic radiation with wavelength λ is,
$$p = h/\lambda \qquad \ldots (3)$$
where h is the Planck's constant. Since the photon has zero rest mass therefore the kinetic energy of the photon ($K = \frac{1}{2}mv^2$) is zero, but still the photon has momentum. Therefore, a body can have a momentum without having kinetic energy

31. The gravitational potential energy is the greatest when the Earth is farthest from the Sun, or when the Northern Hemisphere has summer. (Note that the Earth moves fastest in its orbit, and therefore has the greatest *kinetic energy*, when it is closest to the Sun.)

32. Yes, you can press the upper disk down enough so that when it is released it will spring back and raise the lower disk off the table.
When the upper disk is released after pressing it, the kinetic energy of the upper disk increases until the spring gets its equilibrium length. After that the spring expands and start applying restoring pulling forces on both the disks. When pulling force becomes equal to the weight of the lower disk, it breaks of contact with the table.
Yes, the mechanical energy can be conserved in such a case. In accordance with the law of conservation of mechanical energy, in an isolated system in which only conservative forces act, the total mechanical energy always remains constant. Since there is no external force acting on the system, therefore the total mechanical energy of the system will be conserved.

33. Yes. If the potential energy U is negative (which it can be defined to be), and the absolute value of the potential energy is greater than the kinetic energy K, then the total mechanical energy E will be negative.

34. The work done on the suitcase depends only on (c) the height of the table and (d) the weight of the suitcase.

35. No. Power depends both on the amount of work done by the engine, and the amount of time during which the work is performed. For example, engine 2 will do more work than engine 1, even though it produces half the power, if it operates for more than twice as much time as engine 1.

36. (a) The force is proportional to the negative of the slope of the potential energy curve, so the magnitude of the force will be greatest where the curve is steepest, at point C.

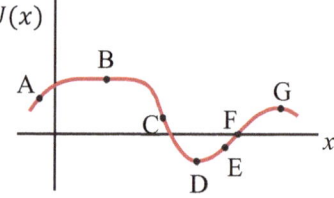

(b) The force acts to the left at points A, E, and F, to the right at point C, and is zero at points B, D, and G. (c) Equilibrium exists at points B, D, and G. B is a point of neutral equilibrium, D is a point of stable equilibrium, and G is a point of unstable equilibrium.

37. (a) If the particle has energy E_3 at x_6, then it has both potential and kinetic energy at that point. As the particle moves toward x_0, it gains kinetic energy as its speed increases. Its speed will be a maximum at x_0. As the particle moves to 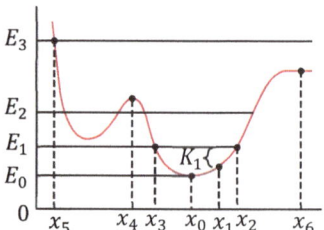 x_4, its speed will decrease, but will be larger than its initial speed. As the particle moves to x_5, its speed will increase, then decrease to zero. The process is reversed on the way back to x_6. At each point on the return trip the speed of the particle is

the same as it was on the forward trip, but the direction of the velocity is opposite. (b) The kinetic energy is greatest at point x_0, and least at x_5.

38. A is a point of unstable equilibrium, B is a point of stable equilibrium, and C is a point of neutral equilibrium.

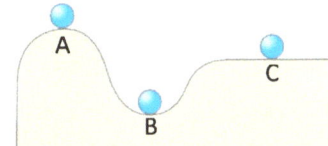

39. Energy is transported from the sun to the earth by electromagnetic waves, such as visible light, that are radiated by the sun. Part of this energy is converted to chemical energy by plants, using photosynthesis. This chemical energy is ingested when we eat the plants or eat an animal that has eaten the plants. Our bodies convert the chemical energy (food energy) into mechanical energy through the operation of our muscles.

33.7. PROBLEMS

1. (a) The force required is $F = ma = (106\ kg)(1.97\ m/S^2) = 209N$. The object moves with an average velocity $v_{av} = v_0/2$ in a time $t = v_0/a$ through a distance $x = v_{av}t = v_0^2/2a$. So,
$x = (51.3m/s)^2/[2(1.97m/s^2)] = 668m$
The work done is, $W = Fx = (-209N)(668m) = 1.40 \times 10^5 J$
(b) The force required is, $F = ma = (106kg)(4.82m/s^2) = 511N$
$x = (51.3m/s)^2/[2(4.82m/s^2)] = 273m$
The work done is, $W = Fx = (-511N)(273m) = 1.40 \times 10^5 J$

2. The force and the displacement are both downwards, so the angle between them is 0°.
$W_G = mgs \cos\theta = (280kg)(9.8ms^{-2})(2.80m) \cos 0°$
$= 7.7 \times 10^3 J$

3. $\sin\theta = 0.902/1.62 \approx 0.6$
and $\cos\theta = \sqrt{1 - \sin^2\theta} = \sqrt{1 - 0.36} = \sqrt{64} = 0.8$
From adjoining figure, we have
$\sin\theta = 0.902/1.62 \approx 0.6$
and $\cos\theta = \sqrt{1 - \sin^2\theta} = \sqrt{1 - 0.36} = \sqrt{64} = 0.8$

Since, block slides down with a constant speed, therefore it is the case of dynamic equilibrium. In this case, we have-
$N = mg \cos\theta$ ... (1)
$F + f_k = mg \sin\theta$ ... (2)

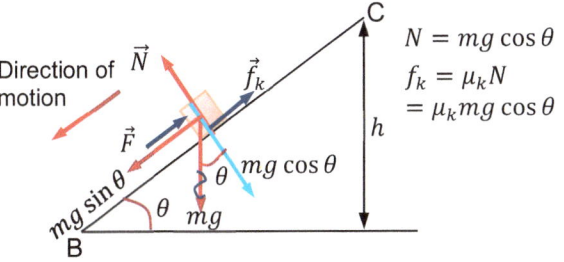

here, $f_k = \mu_k N = \mu_k mg \cos\theta$
Using this value of f_k in equation (2), we get
$F + \mu_k mg \cos\theta = mg \sin\theta$
or $F = mg(\sin\theta - \mu_k \cos\theta)$
$= (47.2)(9.81)[0.6 - 0.11(0.8)]$
$= 214.5$ newton
(a) ∴ the force exerted by the worker,
$F = 214.5$ newton
(b) The work done by the applied force is $W = \vec{F}.\vec{s} = -(214.5\text{newton})(1.62\ m) \approx -347.5J$
(c) The work done by the force of gravity is
$W_{gr} = mgh = (47.2\ kg)(9.81\ m/s^2)(0.902m)$
or $W_{gr} = 417.65J$

4. $F = \dfrac{W}{s} = \dfrac{(4.5e\ )(1.6\times 10^{-1}\ J/eV)}{(3.4\times 10^{-9}m)} = 2.1 \times 10^{-10} N$

5. The rock will rise until gravity does –80.0 J of work on the rock. The displacement is upwards, but the force is downwards, so the angle between them is 180°.
$W_G = mgs \cos\theta$
$s = \dfrac{W_G}{mg\cos\theta} = \dfrac{-80J}{(1.85kg)(9.8ms^{-2})(-1)} = 4.41m$

6. (a) If T is the tension in the cord, then, for the motion of the block
$Mg - T = M(g/4)$
or $T = \dfrac{3Mg}{4}$
(a) The work done by the cord is $W = \vec{F}.\vec{s} = -(3Mg/4)d = -(3/4)Mgd$.
(b) The work done by gravity is $W = \vec{F}.\vec{s} = Mgd$.

7. The incline has a height h where $h = W/mg = (680\ J)/[(75\ kg)(9.81\ m/s^2)] =$. The work required to lift the block is the same regardless of the path, so the length of the incline l is

$l = W/F = 680 \, J/320 \, N$

The angle of the incline is given by

$\sin\theta = \dfrac{h}{l} = \dfrac{F}{mg} = \dfrac{(320N)}{(75kg)(9.81m/s^2)} = \dfrac{320}{735.75} \approx 0.4$

or $\theta \approx 23.6°$

8. **APPROACH:** The gravity force is constant and the displacement is along a straight line, so $W = Fs = \cos\theta$

   The displacement is upward along the ladder and the gravity force is downward, so

   $\theta = 180° - 30° = 150°. w = mg = 735N$

   SOLUTION
   $W = (735N)(2.75m)\cos 150° = -1750$

   (b) No, the gravity force is independent of the motion of the painter.
   - Gravity is downward and the vertical component of the displacement is upward, so the gravity force does negative work.

9. The minimum force required to lift the firefighter is equal to his weight. The force and the displacement are both upwards, so the angle between them is $0°$.

   $W_{\text{climb}} = F_{\text{climb}} s \cos\theta = mgs\cos\theta$
   $= (75.0 \text{kg})(9.80 \text{ms}^{-2})(20.0 \text{m}) \cos 0° = 1.47 \times 10^4 J$

10. The first book is already in position, so no work is required to position it. The second book must be moved upwards by a distance s, by a force equal to its weight, mg. The force and the displacement are in the same direction, so the work is mgs. The third book will need to be moved a distance of $2s$ by the same size force, so the work is $2mgs$. This continues through all seven books, with each needing to be raised by an additional amount of s by a force of mg. The total work done is

    $W = mgs + 2mgs + 3mgs + 4mgs$
    $\qquad + 5mgs + 6mgs + 7mgs$
    $= 28mgs = 28(1.8\text{kg})(9.8\text{ms}^{-2})(0.040\text{m})$
    $= 2.0 \times 10^1 J$

11. **APPROACH:** We want to find the work done by a known force acting through a known displacement.

    $W = \vec{F} \cdot \vec{s} = F_x s_x + F_y s_y$

    We know the components of $\vec{F}$ but need to find the components of the displacement $\vec{s}$.

    SOLUTION Using the magnitude and direction of $\vec{s}$, its components are

    $x = (48\text{m})\cos 240° = -24.0 \text{ m}$ and
    $y = (48\text{m})\sin 240° = -41.57\text{m}$. Therefore,
    $\vec{s} = (-24.0\text{m})\hat{\imath} + (-41.57\text{m})\hat{\jmath}$

    The definition of work gives,

    $W = \vec{F} \cdot \vec{s} = (-68.0N)(-24.0m)$
    $\qquad\qquad + (36.0N)(-41.57m)$
    $\qquad = +1632J - 1497J = +135J$

12. The downward force is $450 \, N$, and the downward displacement would be a diameter of the pedal circle

    $W = Fs \cos\theta = (450N)(0.36m)\cos 0° = 160J$

13. **APPROACH** Calculate the work done by friction and apply $W_{\text{total}} = K_2 - K_1$. Since the friction force is not constant, use Eq. $W = \int f(x)dx$ to calculate the work

    Let x be the distance past P. Since μ_k increases linearly with x, $\mu_k = 0.1 + Ax$. When $x = 12.5 \, m$, $\mu_k = 0.6$, so $A = 0.5/12.5 = 0.04/m$

    SOLUTION (a) $W_{\text{total}} = \Delta K = K_2 - K_1$, gives

    $-\int \mu_k mg\, dx = 0 - \dfrac{1}{2}mv_1^2$

    Using the above expression for μ_k

    $g \int_0^{x_2} (0.100 + Ax)dx = \dfrac{1}{2}v_1^2$ and

    $g\left[(0.100)x_2 + A\dfrac{x_2^2}{2}\right] = \dfrac{1}{2}v_1^2$

    $(9.80\text{ms}^{-2})\left[(0.100)x_2 + (0.04 \, m^{-1})\dfrac{x_2^2}{2}\right]$
    $\qquad = \dfrac{1}{2}(4.5 \text{ ms}^{-1})^2$

    Solving for x_2 gives $x_2 = 5.11 \, m$.

    (b) $\mu_L = 0.100 + (0.0400/m)(5.11m) = 0.304$

    (c) $W_{\text{tot}} = K_2 - K_1$ gives

    $-\mu_2 mg x_2 = 0 - \dfrac{1}{2}mv_1^2 \cdot$

    or $x_2 = \dfrac{v_1^2}{2\mu_1 g} = \dfrac{(4.50\text{m/s})^2}{2(0.100)(9.80\text{m/s}^2)} = 10.3 \, m$

14. The force exerted to stretch a spring is given by $F_{\text{stret}} = kx$

    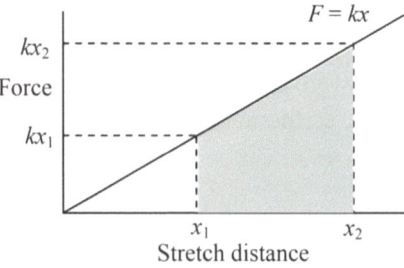

    The opposite of the force exerted by the spring, which is given by $F = -kx$. A graph of F_{stretch} vs. x will be a straight line of slope k through the origin. The stretch from x_1 to x_2, as shown on the graph, outlines a trapezoidal area. This area represents the work

$$W = \tfrac{1}{2}(kx_1 + kx_2)(x_2 - x_1) = \tfrac{1}{2}k(x_1 + x_2)(x_2 - x_1)$$
$$= \tfrac{1}{2}(65 \text{ N/m})(0.095\text{m})(0.035\text{m}) = 0.11\text{J}$$

15. See the graph of force vs. distance. The work done is the area under the graph. It can be found from the formula for a trapezoid

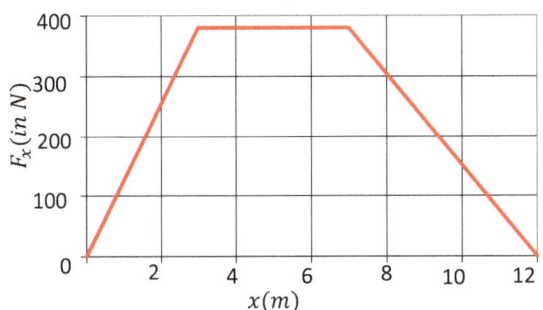

$$W = \tfrac{1}{2}(12.0\text{m} + 4.0\text{m})(380\text{N})$$
$$= 3040\text{J} \approx 3.0 \times 10^3 \text{J}$$

16. The work required to stretch a spring from equilibrium is proportional to the length of stretch, squared. So, if we stretch the spring to 3 times its original distance, a total of 9 times as much work is required for the total stretch. Thus, it would take $45.0\ J$ to stretch the spring to a total of $6.0\ cm$. Since $5.0\ J$ of work was done to stretch the first $2.0\ cm$, $40.0\ J$ of work is required to stretch it the additional $4.0\ cm$.

    This could also be done by calculating the spring constant from the data for the $2.0\ cm$ stretch, and then using that spring constant to find the work done in stretching the extra distance.

17. Treat the area as a trapezoid, with sides of $180\ N$ and $100\ N$, and a base of $20.0\ m$. Then the work is
$$W = \tfrac{1}{2}(20.0\text{m})(180\text{N} + 100\text{N}) \approx 2800\text{J}$$

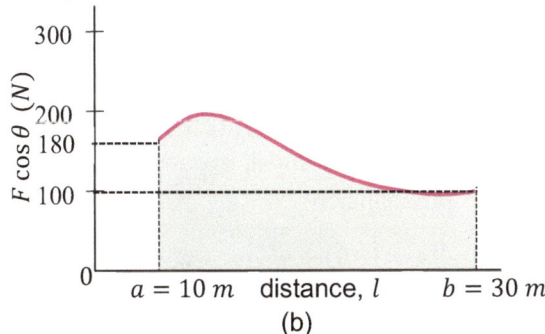

FIGURE Work done by a force F is (a) approximately equal to the sum of the areas of the rectangles, (b) exactly equal to the area under the curve of $F \cos \theta$ vs. l.

18. **APPROACH:** Since velocities at base and top of bridge are provided, therefore, we can easily calculate the total work done on you and bicycle system by using work energy theorem i.e., $W_{tot} = K_2 - K_1$. Neglecting friction, work is done by you (with the force you apply to the pedals) and by gravity i.e., $W_{tot} = W_{you} + W_{grav}$
Therefore, $W_{you} = W_{tot} - W_{grav}$

    SOLUTION Let point 1 be at the base of the bridge and point 2 be at the top of the bridge.
    (a) $W_{tot} = K_2 - K_1$
    $K_1 = \tfrac{1}{2}mv_1^2 = \tfrac{1}{2}(80 \text{ kg})(5 \text{ m/s})^2 = 1000\text{J}$
    $K_2 = \tfrac{1}{2}mv_2^2 = \tfrac{1}{2}(80 \text{ kg})(1.5\text{m/s})^2 = 90\text{J}$
    $W_{tot} = 90 \text{ J} - 1000 \text{ J} = -910 \text{ J}$
    (b) The gravity force is $W = mg = (80\ kg)(9.8\ ms^{-2}) = 784\ N$, downward. The displacement is $5.2\ m$, upward. Thus $\theta = 180°$ and
    $$W_{\text{gravity}} = (F \cos \theta)s = (784\text{N})(5.2\text{m})\cos 180°$$
    $$= -4076.8 \text{ J}$$
    $\therefore W_{you} = W_{tot} - W_{grav}$
    $$= -910 \text{ J} - (-4076.8 \text{ J}) = +3166.8 \text{ J}$$
    ☞ The total work done is negative and you lose kinetic energy.

19. The work done will be the area under the F_x vs. x graph.

    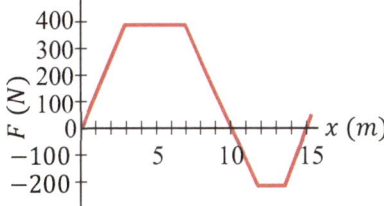

    (a) From $x = 0$ to $x = 10\ m$, the shape under the graph is trapezoidal. The area is
    $$W_a = (400\text{N})\tfrac{1}{2}(10\text{m} + 4\text{m}) = 2800\text{J}$$
    (b) From $x = 10\ m$ to $x = 15\ m$, the force is in the opposite direction from the direction of motion, and so the work will be negative. Again, since the shape is trapezoidal, we find
    $$W_a = (-200\text{N})\tfrac{1}{2}(5\text{m} + 2\text{m}) = -700\text{J}$$
    Thus, the total work from $x = 0$ to $x = 15\ m$
    $$W = 2800\text{J} - 700\text{J} = -2100\text{J}$$

20. Since the force only has an x-component, only the x-displacement is relevant. The object moves from $x = 0$ to $x = d$.
$$W = \int_0^d F_x dx = \int_0^d kx^4 dx = \tfrac{1}{5}kd^5$$

21. **APPROACH** Because the object moves along a straight line, we know that the x-coordinate

increases linearly from 0 to $10\,m$, and the y-coordinate increases linearly from 0 to $20\,m$. Use the relationship, $W = \int_{x_a}^{x_b} F_x dx + \int_{y_a}^{y_b} F_y dy$

SOLUTION $W = \int_{x_a}^{x_b} F_x dx + \int_{y_a}^{y_b} F_y dy$
$= \int_0^{10m} 3.0 x dx + \int_0^{20} 4.0 y dy$

22. $W = \int_0^{3x_0} \vec{F} \cdot d\vec{s} = \frac{F_0}{x_0} \int_0^{3x_0}(x - x_0)\,dx$
$= F_0 x_0 \left(\frac{9}{2} - 3\right)$

23. **APPROACH** The force on the object is given by Newton's law of universal gravitation, $F = G\frac{Mm}{r^2}$ (here M is the mass of earth, m is the mass of space vehicle). The force is a function of distance, so to find the work, we must integrate. The directions are tricky. To use equation. $W = \int \vec{F}.d\vec{s}$, we have $\vec{F} = -G\frac{Mm}{r^2}\hat{r}$ and $d\vec{s} = dr\,\hat{r}$. It is tempting to put a negative sign with the $d\vec{s}$ relationship since the object moves inward, but since r is measured outward away from the center of the Earth, we must not include that negative sign. Note that we move from a large radius to a small radius.

SOLUTION
$W = \int \vec{F}.d\vec{s} = \int_{far}^{near} -G\frac{Mm}{r^2}\hat{r}\cdot(dr\hat{r})$
$= -\int_{R+3300km}^{R} G\frac{Mm}{r^2}\,dr = G\frac{mm_E}{r}\Big|_{R+3300km}^{R}$

(Here, R is the radius of earth)

$= GMm\left(\frac{1}{R} - \frac{1}{R+3300km}\right)$
$= (6.67 \times 10^{-11}\,N.m^2/kg^2)(2800kg)$
$(5.97 \times 10^{24}kg)\left(\frac{1}{6.38\times 10^6 m} - \frac{1}{(6.38+.30)\times 10^6 m}\right)$
$= 6.0 \times 10^{10}\,J$

24. (a) The spring extension is $\delta l = \sqrt{l_0^2 + x^2} - L_0$. The force from one spring has magnitude $k\delta l$, but only the x component contributes to the problem, so,
$F = 2k(\sqrt{l_0^2 + x^2} - l_0)\frac{x}{\sqrt{l_0^2+x^2}}$

is the force required to move the point. The work required is the integral
$W = \int_0^x F dx$ which is
$W = k \cdot x^2 - 2kl_0\sqrt{l_0^2 + x^2} + 2kl_0^2$

Note that it does reduce to the expected behaviour for $x \gg l_0$.

(b) Binomial expansion of square root gives
$\sqrt{l_0^2 + x^2} = l_0\left(1 + \frac{1}{2}\frac{x^2}{l_0^2} - \frac{1}{8}\frac{x^4}{l_0^4}\cdots\right)$,
so, the first term in the above expansion cancels with the last term in W; the second term cancels with the first term in W, leaving

25. Number the springs clockwise from the top of the picture. Then the four forces *on* each spring are

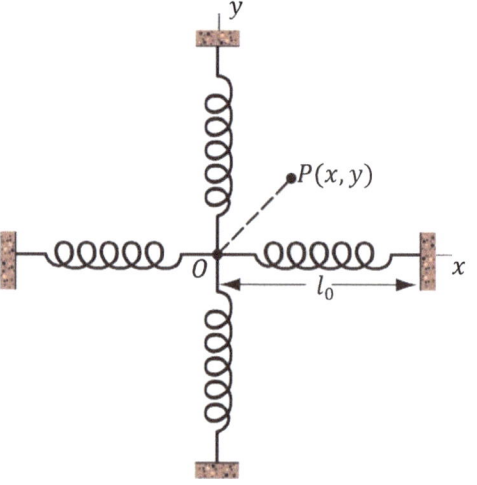

$F_1 = k(l_0 - \sqrt{x^2 + (l_0 - y)^2})$
$F_2 = k(l_0 - \sqrt{(l_0 - x)^2 + y^2})$
$F_3 = k(l_0 - \sqrt{x^2 + (l_0 + y)^2})$
$F_4 = k(l_0 - \sqrt{(l_0 + x)^2 + y^2})$

The directions are *much* harder to work out, but for small x and y we can assume that

$\vec{F}_1 = k(l_0 - \sqrt{x^2 + (l_0 - y)^2})\hat{j}$
$\vec{F}_2 = k(l_0 - \sqrt{(l_0 - x)^2 + y^2})\hat{i}$
$\vec{F}_3 = k(l_0 - \sqrt{x^2 + (l_0 + y)^2})\hat{j}$
$\vec{F}_4 = k(l_0 - \sqrt{(l_0 + x)^2 + y^2})\hat{i}$

Then,
$W = \int \vec{F}\cdot d\vec{s} = \int (F_1 + F_3)dy + \int (F_2 + F_4)dx$

Since x and y are small, expand the force(s) in a binomial expansion:

$F_1(x,y) \approx F_1(0,0) + \frac{\partial F_1}{\partial x}\Big|_{x,y=0} x + \frac{\partial F_1}{\partial y}\Big|_{x,y=0} y = ky$

there will be similar expression for the other four forces. Then
$W = \int 2ky dy + \int 2kx dx = k(x^2 + y^2) = kd^2$

26. **APPROACH** Let y represent the length of chain hanging over the table, and let λ represent the weight per unit length of the chain. Then the force

of gravity (weight) of the hanging chain is $F_G = \lambda y$. As the next small length of chain dy comes over the table edge, gravity does an infinitesimal amount of work on the hanging chain given by the force times the distance, $F_G dy = \lambda y \, dy$. To find the total amount of work that gravity does on the chain, integrate that work expression, with the limits of integration representing the amount of chain hanging over the table.

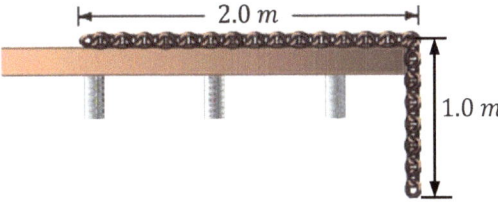

SOLUTION $W = \int_{y_{initial}}^{y_{final}} F_G dy = \int_{1\,m}^{3\,m} \lambda y \, dy$

$= \frac{1}{2}\lambda y^2|_{1.0m}^{3.0m} = \frac{1}{2}(18\,N/m)(9.0m^2 - 1.0m^2) = 72J$

27. **APPROACH** The kinetic energy of the spring would be found by adding together the kinetic energy of each infinitesimal part of the spring. The mass of an infinitesimal part is given by $dm = \frac{M_s}{D}dx$ and the speed of an infinitesimal part is $v = \frac{x}{D}v_0$. Calculate the kinetic energy of the mass + spring system.

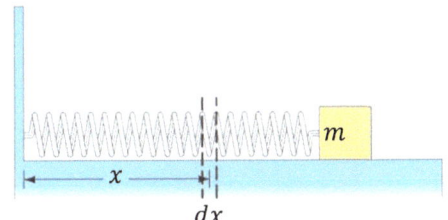

SOLUTION $K = K_{mass} + K_{spring}$

$= \frac{1}{2}mv_0^2 + \frac{1}{2}\int_{mass} v^2 dm$

$= \frac{1}{2}mv_0^2 + \frac{1}{2}\int_0^D \left(v_0\frac{x}{D}\right)^2 \frac{M_s}{D}dx$

$= \frac{1}{2}mv_0^2 + \frac{v_0^2 M_s}{D^3}\frac{1}{2}\int_0^D x^2 dx$

$= \frac{1}{2}mv_0^2 + \frac{1}{2}\frac{v_0^2 M_s}{D^3}\frac{D^3}{3} = \frac{1}{2}v_0^2\left(m + \frac{1}{3}M_s\right)$

So, for a generic speed v, we have,

$K = \frac{1}{2}\left(m + \frac{1}{3}M_s\right)v^2$

or $K = \frac{1}{2}Mv^2$, where, $M = m + \frac{1}{3}M_s$

28. **APPROACH:** In this case, the work is done by the spring and by gravity. Let point A be where the textbook is released and point B be where it stops sliding. $x_B = 0$, since at point B the spring is neither stretched nor compressed. The situation is sketched in adjoining figure.

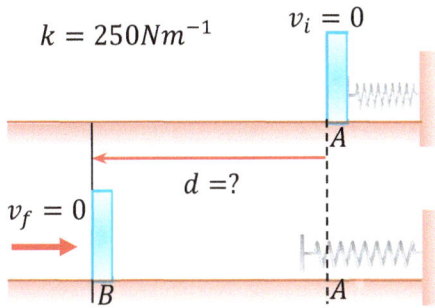

Use work energy theorem $W_{tot} = K_f - K_i$, with $K_i = 0, K_f = 0$ and $W_{tot} = W_{fric} + W_{spr}$

SOLUTION $W_{spr} = \frac{1}{2}kx_A^2$, where $x_A = 0.25\,m$

(Spring force is in direction of motion of block so it does positive work.)

$W_{fric} = -\mu_k mgd$

Then $W_{tot} = K_f - K_i$, gives

$\frac{1}{2}kx_A^2 - \mu_k mgd = 0$

$d = \frac{kx_A^2}{2\mu_k mg} = \frac{(250N/m)(0.250m)^2}{2(0.30)(2.50kg)(9.80\,ms^{-2})} = 1.1m$

measured from the point where the block was released.

☞ The positive work done by the spring equals the magnitude of the negative work done by friction. The total work done during the motion between points A and B is zero and the textbook starts and ends with zero kinetic energy.

29. **APPROACH:** Apply $W_{tot} = K_2 - K_1$, with $W = Fs\cos\theta$

The students do positive work, and the force that they exert makes an angle of $30°$ with the direction of motion. Gravity does negative work, and is at an angle of $120°$ with the chair's motion.

SOLUTION: The total work done is-

$W_{tot} = [(600N)\cos 30°$
$\qquad +(85kg)(9.8ms^{-2})\cos 120°](2.5m)$
$\qquad = 257.8J$

and so, the speed at the top of the ramp is-

$v_2 = \sqrt{v_1^2 + \frac{2W_{tot}}{m}} = \sqrt{(2\,m/s)^2 + \frac{2(257.8J)}{(85kg)}}$

$\qquad = 3.17\,m/s$

☞ The component of gravity down the incline is $mg\sin 30° = 417N$ and the component of the push up the incline is $(600\,N)\cos 30° = 520\,N$. The force component up the incline is greater than the force component down the incline; the net work done is positive and the speed increases

30. (a) $W_g = -(0.263\text{kg})(9.81 ms^{-2})(-0.118 m)$
$= 0.304 J$

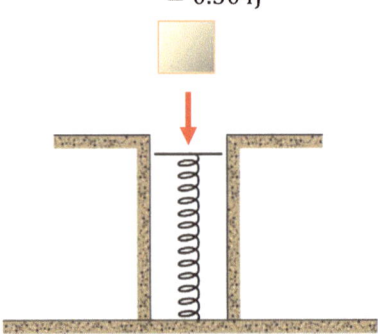

(b) $W_s = -\frac{1}{2}(252 N/m)(-0.118 m)^2 = -1.75 J$

(c) The kinetic energy just before hitting the block would be $1.75 J - 0.304 J = 1.45 J$. The speed is then $v = \sqrt{2(1.45 J)/(0.263\text{kg})} = 3.32 m/s$

(d) Doubling the speed quadruples the initial kinetic energy to $5.78 J$. The compression will then be given by

$-5.78 J = -\frac{1}{2}(252 N/m)y^2 - (0.263\text{kg})(9.81 m/s^2)y$

with solution $y = 0.225\ m$

31. (a) We can solve this with a trick of integration.

$W = \int_0^x F dx = \int_0^x m a_x \frac{dt}{dt} dx = m a_x \int_0^t \frac{dx}{dt} dt$
$= m a_x \int_0^t v_x dt = m a_x \int_0^t a t dt = \frac{1}{2} m a_x^2 t^2$

Basically, we changed the variable of integration from x to t, and then used the fact the the acceleration was constant so $v_x = v_{0x} + a_x t$. The object started at rest so $v_{0x} = 0$, and we are given in the problem that $v_f = a t_f$. Combining,

$W = \frac{1}{2} m a_x^2 t^2 = \frac{1}{2} m \left(\frac{v_f}{t_f}\right)^2 t^2$

(b) Instantaneous power will be the derivative of this, so,

$P = \frac{dW}{dt} = m \left(\frac{v_f}{t_f}\right)^2 t$

32. Let M be the mass of the helicopter. It will take a force Mg to keep the helicopter airborne. This force comes from pushing the air down at a rate $\Delta m / \Delta t$ with a speed of v; so $Mg = v \Delta m / \Delta t$. The blades sweep out a cylinder of cross sectional area A, so $\Delta m / \Delta t = \rho A v$. The force is then $Mg = \rho A v^2$; the speed that the air must be pushed down is $v = \sqrt{Mg/\rho A}$. The minimum power is then

$P = F v = M g \sqrt{\frac{Mg}{\rho A}} = \sqrt{\frac{(1820\text{kg})^3 (9.81 m^{-2})^3}{(1.23\text{kg}/m^3)\pi(4.88 m)^2}}$
$= 2.49 \times 10^5 W$

33. **APPROACH:** Once the block leaves the top of the hill it moves in projectile motion. Use Eq. (7.14) to relate the speed v_B at the bottom of the hill to the speed v_{Top} at the top and the $70\ m$ height of the hill.

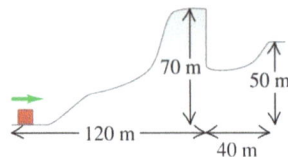

For the projectile motion, take $+y$ to be downward.
$a_x = 0, a_y = g, v_{0x} = v_{Top}, v_{0y} = 0$
For the motion up the hill only gravity does work. Take $y = 0$ at the base of the hill.

SOLUTION First get speed at the top of the hill for the block to clear the pit $y = \frac{1}{2} g t^2$

$20 m = \frac{1}{2}(9.8 m/s^2) t^2$, $t = 2.0 s$.

Then $v_{Top} t = 40 m$ gives $v_{Top} = \frac{40}{2.0 s} = 20 m/s$

By conservation of energy, we have

$K_{Bottom} = U_{Top} + K_{Top}$
$\Rightarrow \frac{1}{2} m v_B^2 = mgh + \frac{1}{2} m v_{Top}^2$

or $v_B = \sqrt{v_{Top}^2 + 2gh}$
$= \sqrt{(20 m/s)^2 + 2(9.8 m/s^2)(70 m)} = 42 m/s$

34. There are two forces on the woman, the force of gravity directed down and the normal force of the floor directed up. These will be effectively equal, so $N = W = mg$. Consequently, the $57 kg$ woman must exert a force of $F = (57\text{kg})(9.8 ms^{-2})$ to propel herself up the stairs.

From the reference frame of the woman the stairs are moving down, and she is exerting a force down, so the work done by the woman is given by

$W = Fs = (560\ N)(4.5\ m) = 2500 J$ this work is positive because the force is in the same direction as the displacement. The average power supplied by the woman is given by

$P = \frac{W}{t} = \frac{2500 J}{3.5 s} = 710\ W$

35. $P = \vec{F} \cdot \vec{v} = (720\ N)(26\ ms^{-1}) = 19000\ W$; in horsepower, $P = 19000\ W = \frac{19000}{745.7} hp = 25 hp$

36. $P = 6.6 \times 745.7 = 4921.62\ W$, and the flow rate is $Q = 220\ gal/min = \frac{220 \times 3.785}{60} litre/s = 13.9\ litre/s$
    $\because$ the density of water $\rho \approx 1.0\ kg/L$,
    $\therefore$ the mass flow rate is $R = \rho Q = 13.9\ kg/s$

37. Force and instantaneous power are related by
    $$P = Fv$$
    or $\quad F = \frac{P}{v}$
    or $\quad m\frac{dv}{dt} = \frac{P}{v}\ [F = ma = m\frac{dv}{dt}]$
    or $\quad m\frac{dx}{dt}\frac{dv}{dx} = \frac{P}{v}$
    or $\quad mv\frac{dv}{dx} = \frac{P}{v}$
    or $\quad \int_0^v mv^2 dv = \int_0^x P dx$
    or $\quad \frac{1}{3}mv^3 = Px$
    or $\quad v = (3xP/m)^{1/3}$

38. (a) If the drag is $D = bv^2$, then the force required to move the plane forward at constant speed is $F = D = bv^2$, so the power required is $P = Fv = bv^3$
    (b) $P \propto v^3$, so if the speed increases to 125 % then P increases by a factor of $1.25^3 = 1.953$, or increases by 95.3 %

39. (a) $dP/dv = ab - 3av^2$, so P_{max} occurs when $3v^2 = b$, or $v = \sqrt{b/3}$
    (b) $F = P/v$, so $dF/dv = -2v$, which means F is a maximum when $v = 0$
    (c) No; $P = 0$, but $F = ab$

40. $U(x) - U(x_0) = -\int_{x_0}^x F_x(x) dx$
    $= -\int_{x_0}^x (-\alpha x e^{-\beta x^2}) dx - \frac{-\alpha}{2\beta} e^{-\beta x^2} \Big|_{x_0}^x$
    or $U(x) = U(x_0) + \frac{\alpha}{2\beta}(e^{-\beta x_0^2} - e^{-\beta x^2})$
    If we choose $x_0 = 1$ and $U(x_0) = 0$, then
    $$U(x) = -\frac{\alpha}{2\beta} e^{-\beta x^2}$$

41. (a) Define the system as the block and the Earth.

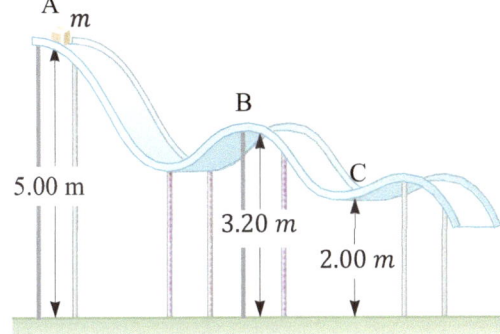

$\Delta K + \Delta U = 0$
or $\left(\frac{1}{2}mv_B^2 - 0\right) + (mgh_B - mgh_A) = 0$
or $\frac{1}{2}mv_B^2 = mg(h_A - h_B)$
or $v_B = \sqrt{2g(h_A - h_B)}$
or $v_B = \sqrt{2(9.80 ms^{-2})(5.00m - 3.20m)}$
$\qquad = 5.94 m/s$
or $v_c = \sqrt{2g(h_A - h_c)}$
or $v_c = \sqrt{2g(5.00 - 2.00)} = 7.67 m/s$
(b) Treating the block as the system,
$W_g|_{A \to C} = \Delta K = \frac{1}{2}mv_C^2 - 0$
$= \frac{1}{2}(5.00 kg)(7.67 m/s)^2 = 147 J$

42. (a) Since $\Delta y = 0$, then $\Delta U = 0$ and $\Delta K = 0$. Consequently, at B, $v = v_0$.

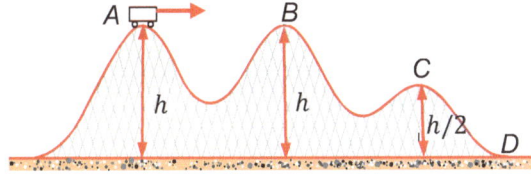

(b) At C $K_C = K_A + U_A - U_C$,
or $\frac{1}{2}mv_C^2 = \frac{1}{2}mv_0^2 + mgh - mg\frac{h}{2}$
or $v_B = \sqrt{v_0^2 + 2g\frac{h}{2}} = \sqrt{v_0^2 + gh}$
(c) At D $K_D = K_A + U_A - U_D$,
or $\frac{1}{2}mv_D^2 = \frac{1}{2}mv_0^2 + mgh - mg(0)$
or $v_B = \sqrt{v_0^2 + 2gh}$

43. When block B moves up by $1\ cm$ block A moves down by $2\ cm$ and the separation becomes $3\ cm$. We then choose the final point to be when B has moved up by $h/3$ and has speed $v_A/2$. Then A has moved down $2h/3$ and has speed v_A.
$\Delta K + \Delta U = 0$
or $(K_A + K_B + U_g)_f - (K_A + K_B + U_g)_i = 0$
or $(K_A + K_B + U_g)_i = (K_A + K_B + U_g)_f$
or $0 + 0 + 0 = \frac{1}{2}mv_A^2 +$
$\frac{1}{2}m\left(\frac{v_A}{2}\right)^2 + \frac{mgh}{3} - \frac{mg2h}{3}$
or $\frac{mgh}{3} = \frac{5}{8}mv_A^2$
or $v_A = \sqrt{\frac{8gh}{15}}$

MECHANICS

44. (a) The force constant of the spring is
$k = \frac{F}{x} = \frac{mg}{x} = \frac{(7.94\text{kg})(9.81 ms^{-2})}{0.102 m} =$ 764N/m

(b) The potential energy stored in the spring is given by-
$U = \frac{1}{2}kx^2 = \frac{1}{2}(764\text{ N/m})(0.286m + 0.102m)^2$
$= 57.5\text{ J}$

(c) Conservation of energy,
$K_f + U_f = K_i + U_i$
or $\frac{1}{2}mv_f^2 + mgy_f + \frac{1}{2}kx_f^2 = \frac{1}{2}mv_i^2 + mgy_i + \frac{1}{2}kx_i^2$
or $\frac{1}{2}(0)^2 + mgh + \frac{1}{2}k(0)^2 = \frac{1}{2}(0)^2 + mg(0) + \frac{1}{2}kx_i^2$

Rearranging,
$h = \frac{k}{2m}x_i^2 = \frac{(764 \text{ /m})}{2(7.94\text{kg})(9.81 \text{ /s}^2)}(0.388m)^2 = 0.738m$

45. Let the spring get compressed a distance x. If the object fell from a height $h = 0.436\ m$, then conservation of energy gives

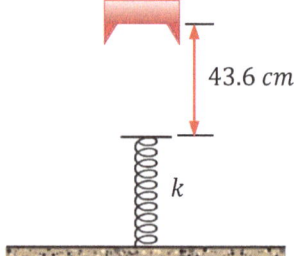

43.6 cm

$\frac{1}{2}kx^2 = mg(x + h)$

Solving for x, we get
$x = \frac{mg}{k} \pm \sqrt{\left(\frac{mg}{k}\right)^2 + 2\frac{mg}{k}h}$

only the positive answer is of interest, so
$x = \frac{(2.14\text{kg})\left(\frac{9.81m}{s^2}\right)}{\left(\frac{1860\text{N}}{m}\right)}$
$\pm \sqrt{\left(\frac{(2.14\text{kg})(9.81ms^{-2})}{(1860\ Nm^{-1})}\right)^2 + 2\frac{(2.14\text{kg})((9.81ms^{-2}))}{(1860\ Nm^{-1})}(0.436m)}$
$= 0.111m$

46. (a) Apply conservation of energy to the bead-string-Earth system to find the speed of the bead at B. Friction transforms mechanical energy of the system into internal energy $\Delta E_{int} = f_k d$.

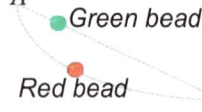

$\Delta K + \Delta U + \Delta E_{int} = 0$
$\left[\frac{1}{2}mv_B^2 - \frac{1}{2}mv_A^2\right] + (mgy_B - mgy_A) + f_k d = 0$

or $\left[\frac{1}{2}mv_B^2 - 0\right] + (0 - mgy_A) + f_k d = 0$
or $\frac{1}{2}mv_B^2 = mgy_A - f_k d$
or $v_B = \sqrt{2gy_A - \frac{2f_k d}{m}}$

For, $y_A = 0.200m, f_k = 0.025N, d = 0.600m$, and $m = 25.0 \times 10^{-3}\text{kg}$:

$v_B = \sqrt{2(9.80 \text{ m/s}^2)(0.200m) - \frac{2(0.025N)(0.600m)}{25.0 \times 10^{-3}\text{kg}}}$
$= \sqrt{2.72}\text{m/s}$
$v_B = 1.65\text{m/s}$

47. (a) Yes, the child-Earth system is isolated because the only force that can do work on the child is her weight. The normal force from the slide can do no work because it is always perpendicular to her displacement. The slide is frictionless, and we ignore air resistance.

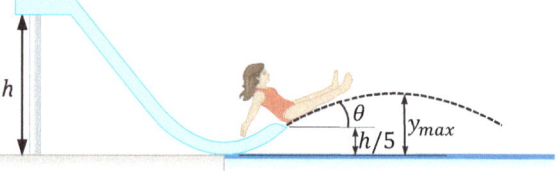

(b) No, because there is no friction.

(c) At the top of the water slide,
$U_g = mgh$ and $K = 0: E = 0 + mgh$
$E = mgh$

(d) At the launch point, her speed is v_i, and height $h = h/5$: $\quad E = K + U_g$
or $E = \left[\frac{1}{2}mv_i^2 + \frac{mgh}{5}\right]$

(e) At her maximum airborne height, $h = y_{max}$:
or $E = \frac{1}{2}mv^2 + mgh = \frac{1}{2}m(v_{xi}^2 + v_{yi}^2) + mgy_{max}$
or $E = \frac{1}{2}m(v_{xi}^2 + 0) + mgy_{max}$
or $E = \left[\frac{1}{2}mv_{xi}^2 + mgy_{max}\right]$

(g) At the launch point, her velocity has components $v_{xi} = v_i \cos\theta$ and $v_{yi} = v_i \sin\theta$:
$E = \frac{1}{2}mv_i^2 + \frac{mgh}{5} = \frac{1}{2}mv_{xi}^2 + mgy_{max}$
or $\frac{1}{2}mv_i^2 + \frac{mgh}{5} = \frac{1}{2}m(v_i \cos\theta)^2 + mgy_{max}$
or $\frac{1}{2}v_i^2(1 - \cos^2\theta) + \frac{gh}{5} = gh_{max}$
or $h_{max} = \frac{1}{2g}\left(\frac{8gh}{5}\right)(1 - \cos^2\theta) + \frac{gh}{5g}$
or $h_{max} = \left(\frac{4h}{5}\right)(1 - \cos^2\theta) + \frac{h}{5}$
or $h_{max} = h(1 - \frac{4}{5}\cos^2\theta)$

(h) No. If friction is present, mechanical energy of the system would *not* be conserved, so her kinetic energy at all points after leaving the top of the waterslide would be reduced when compared with the frictionless case. Consequently, her launch speed, maximum height reached, and final speed would be reduced as well.

48. **APPROACH** Assume the spring to be ideal that obeys Hooke's law, and model the block as a particle.

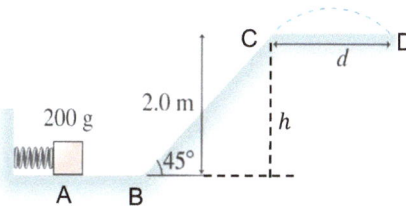

We place the origin of the coordinate system on the free end of the compressed spring which is in contact with the block (position A). Because the horizontal surface at the bottom of the ramp is frictionless, the spring energy appears as kinetic energy of the block until the block begins to climb up the incline.

When block reaches to top of incline (position C), block's mechanical energy decreases (converts to thermal energy) due to non-conservative force of kinetic friction. So, by using relation W_{nc} = change in mechanical energy between B and C), we can calculate the kinetic energy and hence the velocity of the block at the top of incline. Now, to find d, apply the formula of range in projectile motion for same level projection with launch angle $\theta_0 = 45°$

SOLUTION Applying energy conservation between positions A and B of the block

$E_A = E_B$

$\frac{1}{2}kx^2 = K_B$ ... (1)

Let length $BC = s = 2m$, and f_k is the force of kinetic friction, then between points B and C, work done by friction = change in mechanical energy
i.e., $-f_k s = E_C - E_B = mgh + K_C - K_B$
(Here, we have considering AB as reference gravitational potential energy level.)

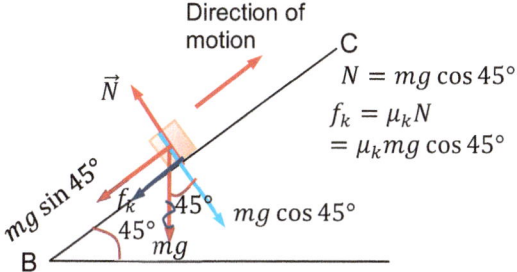

$N = mg \cos 45°$
$f_k = \mu_k N = \mu_k mg \cos 45°$

or $K_C = K_B - mgh - f_k s = \frac{1}{2}kx^2 - mgh - f_k s$

($\because$ from equation (1), $K_B = \frac{1}{2}kx^2$)

or $K_C = \frac{1}{2}(1000 \ N/m)(0.15m)^2$
$-(0.2kg)(9.8 \ m/s^2)(2m \sin 45°)$
$-(0.2)(0.2 \ kg)(9.8 \ m/s^2 \cos 45°)(2m)$
($\because f_k = \mu_k N = \mu_k mg \cos 45°$)
$\Rightarrow K_C = 11.25 \ N.m - 2.77 \ kg.m^2/s^2$
$-0.55 \ kg.m^2/s^2$

or $\frac{1}{2}mv_C^2 = 7.93 J$ or $v_C = \sqrt{\frac{2(7.93J)}{0.2 \ kg}} = 8.9 m/s$

$d = \frac{v_2^2 \sin 2\theta_0}{g} = \frac{(8.9 \ m/s)^2 \sin 2(45°)}{9.8 m/s^2} \approx 8.1 \ m$

49. The horizontal distance travelled by the marble is $R = v t_f$, where t_f is the time of flight and v is the speed of the marble when it leaves the gun. We find *that* speed using energy conservation principles applied to the spring just before it is released and just after the marble leaves the gun.

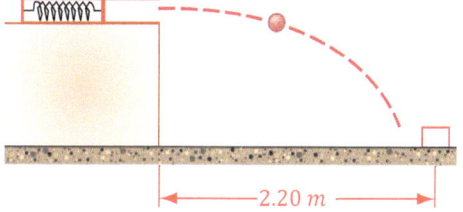

$K_i + U_i = K_f + U_f$

or $0 + \frac{1}{2}kx^2 = \frac{1}{2}mv^2 + 0$

$K_i = 0$ because the marble isn't moving originally, and $U_f = 0$ because the spring is no longer compressed. Substituting R into this,

$\frac{1}{2}kx^2 = \frac{1}{2}m\left(\frac{R}{t_f}\right)^2$

We have two values for the compression, x_1 and x_2, and two ranges, R_1 and R_2. We can put both pairs into the above equation and get two expressions; if we divide one expression by the other, we get

$\left(\frac{x_2}{x_1}\right)^2 = \left(\frac{R_2}{R_1}\right)^2$

We can easily take the square root of both sides, then

$\frac{x_2}{x_1} = \frac{R_2}{R_1}$

R_1 was Manoj's try, and was equal to $2.20 - 0.27 = 1.93 \ m$. $x_1 = 1.1 \ cm$ was his compression. If Tony wants to score, she wants $R_2 = 2.2 \ m$, then

$x_2 = \frac{2.2m}{1.93m} 1.1 cm = 1.25 cm$

50. $P\Delta t = W = \Delta K = \frac{(\Delta m)v^2}{2}$

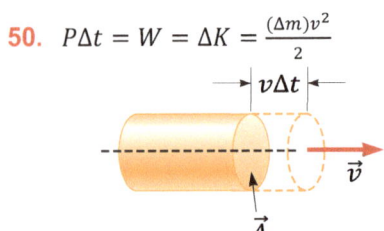

The density is, $\rho = \frac{\Delta m}{\text{volume}} = \frac{\Delta m}{A\,\Delta x}$

Substituting this into the first equation and solving for P, since $\frac{\Delta x}{\Delta t} = v$, for a constant speed, we get

$P = \sqrt{\frac{\rho A\, ^3}{2}}$

Also, since $P = Fv$,

$F = \frac{\rho A v^2}{2}$

Our model predicts the same proportionalities as the empirical equation, and gives $D = 1$ for the drag coefficient. Air actually slips around the moving object, instead of accumulating in front of it. For this reason, the drag coefficient is not necessarily unity. It is typically less than one for a streamlined object and can be greater than one if the airflow around the object is complicated.

51. (a) The energy stored in the spring is the elastic potential energy,

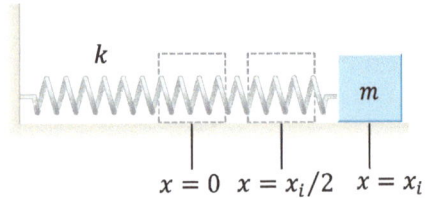

$U = \frac{1}{2}kx^2$, where $k = 850\ N/m$. At $x = 6.00\ cm$,

$U = \frac{1}{2}kx^2 = \frac{1}{2}(850\ \text{N/m})(0.0600\text{m})^2 = 1.53 J$

At the equilibrium position, $x = 0$, $U = 0 J$.

(b) Applying energy conservation to the block-spring system: $\Delta K + \Delta U = 0$

$\left(\frac{1}{2}mv_f^2 - \frac{1}{2}mv_i^2\right) + (U_f - U_i) = 0$

or $\left(\frac{1}{2}mv_f^2 - 0\right) = -(U_f - U_i)$

or $\frac{1}{2}mv_f^2 = U_i - U_t$

52. (a) Let mass m_1 of the chain laying on the table and mass m_2 hanging off the edge. For the hanging part of the chain, apply the particle in equilibrium model in the vertical direction:

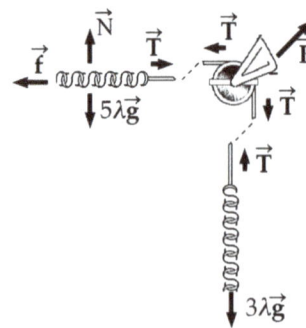

$m_2 g - T = 0$ ... (1)

For the part of the chain on the table, apply the particle in equilibrium model in both directions:

$N - m_1 g = 0$ ... (2)
$T - f_s = 0$ ... (3)

Assume that the length of chain hanging over the edge is such that the chain is on the verge of slipping. Add equations (1) and (3), impose the assumption of impending motion, and substitute equation (2):

$N - m_1 g = 0$

or $f_s = m_2 g$

or $\mu_s N = m_2 g$

or $\mu_s m_1 g = m_2 g$

or $m_2 = \mu_s m_1 = 0.6 m_1$

From the total length of the chain of 8.00 m, we see that $m_1 + m_2 = 8\lambda$

where λ is the mass of a one meter length of chain. Substituting for m_2

$m_1 + 0.6 m_1 = 8\lambda$

or $1.60\, m_1 = 8\lambda$

or $m_1 = 5\lambda$

From this result, we find that $m_2 = 3\lambda$ and we see that 3m of chain hangs off the table in the case of impending motion.

(b) Let x represent the variable distance the chain has slipped since the start.

Then length $(5 - x)$ remains on the table, with now $\Sigma F_y = 0$: $N - (5 - x)\lambda g = 0$

or $N = (5 - x)\lambda g$

or $f_k = \mu_k N = 0.4(5 - x)\lambda g = 2\lambda g - 0.4 \times \lambda g$

Consider energies of the chain-Earth system at the initial moment when the chain starts to slip, and a final moment when $x = 5$, the last link goes over the brink. Measure heights above the final position of the leading end of the chain. At the moment the final link slips off, the center of the chain is at $y_f = 4$ meters.

Originally, 5 meters of chain is at height $8\,m$ and the middle of the dangling segment is at height $8 - \frac{3}{2} = 6.5\,m$.

$K_1 + U_1 + \Delta E_{mech} = K_1 + U_1$:

$0 + (m_1 g y_1 + m_2 g y_2)_1 - \int_i^1 f_k dx$
$\qquad = \left(\frac{1}{2}mv^2 + mgy\right)_t$

$(5\lambda g)8 + (3\lambda g)6.5 - \int_0^5 (2\lambda g - 0.4 \times \lambda g)dx$
$\qquad = \frac{1}{2}(8\lambda)v^2 + (8\lambda g)4$

$40g + 19.5g - 2.00g\int_0^5 dx + 0.4g\int_0^5 x\,dx$
$\qquad = 4v^2 + 32g$

$27.5g - 2gx|_0^5 + 0.4g\frac{x^2}{2}\Big|_0^5 = 4.00v^2$

$27.5g - 2.00g(5) + 0.4g(12.5) = 4.00v^2$

$22.5g = 4.00v^2$

$v = \sqrt{\frac{(22.5m)(9.80\,ms^{-2})}{4}} = 7.42\,ms^{-1}$

53. $F = -\partial U/\partial r$,

or $F = -U_0\left(\frac{r_0}{r^2} + \frac{1}{r}\right)e^{-r/r_0}$

(b) Evaluate the force at the four points:

$F(r_0) = -2(U_0/r_0)e^{-1}$
or $F(2r_0) = -(3/4)(U_0/r_0)e^{-2}$
or $F(4r_0) = -(5/16)(U_0/r_0)e^{-4}$
or $F(10r_0) = -(11/100)(U_0/r_0)e^{-10}$

The ratios are then

$F(2r_0)/F(r_0) = (3/8)e^{-1} = 0.14$
or $F(4r_0)/F(r_0) = (5/32)e^{-3} = 7.8 \times 10^{-3}$
or $F(10r_0)/F(r_0) = (11/200)e^{-9} = 6.8 \times 10^{-6}$

33.8. MULTIPLE CHOICE PROBLEMS

1. (A) Work done by conservative force $= -\Delta U =$ positive $\Rightarrow \Delta U \downarrow$

2. (B) Average power
$$s^\wedge P_{av} = \frac{\text{Network done}}{\text{Total time taken}}$$
Net work done = change in kinetic energy
= Final energy – initial energy
$= \frac{1}{2} \times 10 \times 2^2 = 20J$ ... (1)
Average power $= \frac{20}{20} = 1$ watt

3. (B) By conservation of mechanical energy, we have
$= \frac{1}{2}kx^2$
or $x = \frac{2mg}{k}$

4. (A) Since mg & N are perpendicular to velocity $\vec{v}$ and $d\vec{s}$, therefore the work done by these forces is zero. Since no relative sliding occurs during walking, static friction comes into play. As the point of application of static frictional force does not move in ground frame, therefore, the work done by static friction in ground frame is zero.

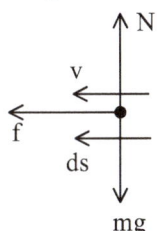

As the man loses his body's internal energy, therefore he performs work. If ΔE is the loss in internal energy of the man, then work performed by him
$W = \Delta E$ (numerically).

5. (C)

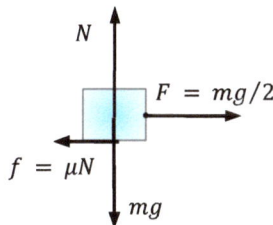

$W = \int dW = \int F dx - \int \mu N dx$
$= \frac{mg}{2}x - mg\mu_0 \int_0^x x dx = 0 \Rightarrow x = \frac{1}{\mu_0}$

6. (D) Power $P = \vec{F}.\vec{v}$,
where F = force imparted by the machine, v = velocity of the liquid
But, Force, F = pressure × area = pA
where p = pressure & A = effective area
$\therefore P = pAv = pA\frac{dx}{dt} = p\frac{dV}{dt}$ ... (1)
Here, $\frac{dV}{dt} = A\frac{dx}{dt}$ = rate of flow of liquid.
Substituting the given values in (1), we get
$\therefore \qquad P = (1.5 \times 10^5)\left(\frac{60}{60} \times 10^{-6}\right) = 0.15\,watt$
$(\because 1\,atm \approx 10^5\,N/m^2)$

Thus, no option is correct.

7. **APPROACH** Since, the spring force and gravitational force both are conservative in nature, therefore, we can apply the principle of conservation of mechanical energy between the positions A and B.

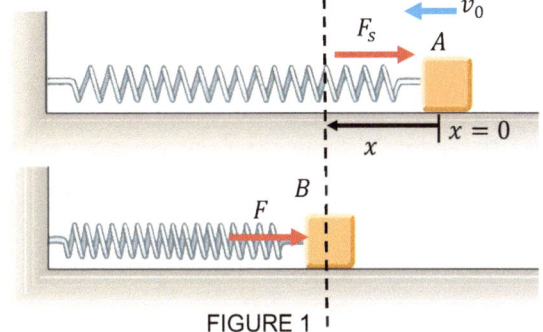

FIGURE 1

98 MECHANICS

SOLUTION (D) By conservation of mechanical energy of spring block system between positions A and B, we have
Total mechanical energy of the system at configuration A = Total mechanical energy of the system at configuration B.
i.e., $(ME)_E = (ME)_B$
or $(KE + U_{gr} + U_{el})_A = (KE + U_{gr} + U_{el})_B$ ... (1)
Here KE, U_{gr} and U_{el} represents kinetic energy, gravitational potential energy and elastic potential energy of the spring block and earth system.
Since, the block's initial and final positions are in same horizontal plane, therefore the gravitational PE will not change with position. So, we can cancel U_{gr} from both sides of the above equation. In this case the Eq. 1, takes the form
$(KE + U_{el})_A = (KE + U_{el})_B$ ... (2)
At position A, the spring is relaxed and speed of the block was v_0, therefore at A, $KE = \frac{1}{2}mv_0^2$, $U_{el} = 0$ whereas, at position B, the block is at rest and the compression in spring is maximum, therefore, $KE = 0$ and $U_{el} = \frac{1}{2}kx^2$
Substituting these values in Eq. 2, we get
or $\frac{1}{2}mv_0^2 + 0 = 0 + \frac{1}{2}kx^2$

or $x = \sqrt{\frac{m}{k}} v_0$

At maximum compression, the spring force on the block will be maximum. It value is given by-
$\therefore \quad F_{max} = kx = k\sqrt{\frac{m}{k}} v_0 = \sqrt{km}\, v_0$

or $F_{max} = \sqrt{km}\, v_0$
$\Rightarrow F_{max} \propto \sqrt{k}$, $F_{max} \propto \sqrt{m}$ and $F \propto v_0$

8. **(D)** Instantaneous power delivered
$P = \vec{F}.\vec{v} = Fv$ where,
$F - f = ma$
$\Rightarrow F = f + ma$
$\Rightarrow P = (f + ma)v$
$\because \quad f = \mu mg$
$\therefore \quad P = (\mu mg + ma)v = m(a + \mu g).at$

9. **(C)** As, gravitational force is a conservative force, therefore, the work done by gravity will depend only upon the height 'h' and it is path independent. So, work done in both the vases $W = mgh$ only.
Therefore, the required ratio $= 1:1$

10. **(A)** Potential energy stored in the spring corresponding to compression x is given by,

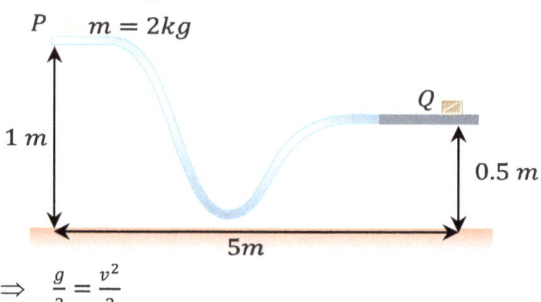

$U = \frac{1}{2}kx^2$

and the magnitude of the spring force, $F = kx$
Therefore, we can write, $U = \frac{1}{2k}k^2x^2 = \frac{F^2}{2k}$

Therefore, the maximum PE, will be given by-
$U_{max} = \frac{F_{max}^2}{2k}$
Now, for limiting equilibrium of the spring mass system, we have
$F_{max} = \mu_s N = \mu mg$
$\therefore \quad U_{max} = \frac{F_{max}^2}{2k} = \frac{(\mu mg)^2}{2k} = \frac{\mu^2 m^2 g^2}{2k}$

11. **(C)** Work done by the external force $= \Delta E$ of the earth-object system $= \Delta PE$ ($\because \Delta KE = 0$)

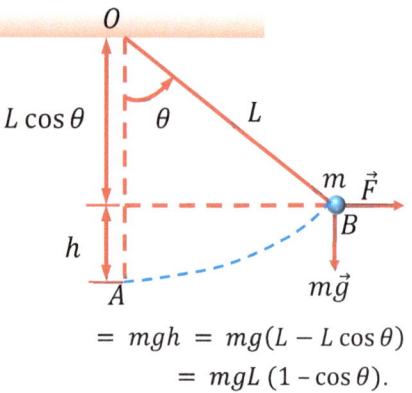

$= mgh = mg(L - L\cos\theta)$
$= mgL(1 - \cos\theta)$.

12. **(A)** Work done against friction $= -$ work done by friction $= \mu mg \cos\theta\, s$.

13. **(D)**

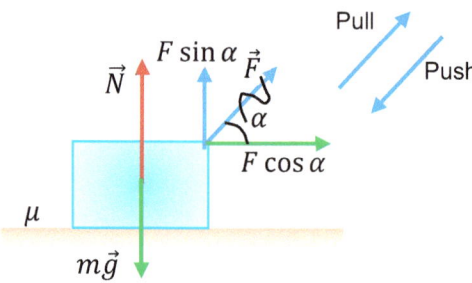

$f_r = \mu(mg - F\sin\alpha)$ pulling
$f_r = \mu(mg + F\sin\alpha)$ pushing.

14. **(D)** $W = \int \vec{F}.d\vec{s} = \int_{s_1}^{s} \frac{k}{s} ds = k \ln s/s_1$

15. **(C)** $mg\left(1 - \frac{1}{2}\right) = \frac{1}{2}mv^2$

$\Rightarrow \frac{g}{2} = \frac{v^2}{2}$

WORK, ENERGY AND POWER

$\Rightarrow v = \sqrt{g} \cong 3.13 \, m/s$.

16. **(A)** The acceleration of the particle is given by
$$a = \frac{F}{m} = \frac{40}{5} = 8 \, m/s^2$$
Velocity at the end of 5 seconds is
$$v = u + at$$
$$v = 0 + 8 \times 5 = 40 \, m/s$$
∴ K.E. acquired $= \frac{1}{2}mv^2 = \frac{1}{2} \times 5 \times 40^2 = 4000 \, J$

17. **(B)** Here displacement is only along z-axis
Therefore, $\vec{s} = 5\hat{k}$ and $\vec{F} = (2\hat{\imath} + 3\hat{\jmath} + 4\hat{k})$
∴ Work done by force $W = \vec{F}.\vec{s}$
$$= (2\hat{\imath} + 3\hat{\jmath} + 4\hat{k}).5\hat{k} = 20J$$

18. **(D)** Work done $dw = \vec{F}.d\vec{s}$

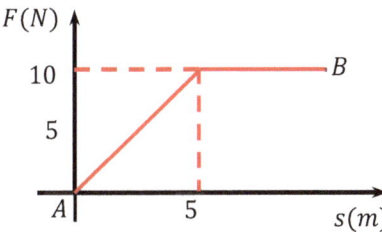

∴ $W = \int \vec{F}.d\vec{s}$ = Area under F-s curve
$= \frac{1}{2} \times 5 \times 10 + 2 \times 10 = 45 \, J$

19. **(D)** ΔK.E. of particle on reaching the
ground $= \frac{1}{2}m.(1.21)gh = 0.605 \, mgH$

By work energy theorem, we have
$W_{net} = K_2 - K_1$
or $W_c + W_{nc} = K_2 - K_1$ ... (1)
Here, W_c = work done by conservative gravitational force = – change in PE
$= -(0 - mgh) = mgh$
Substituting this value of W_c, in (1), we get-
$$mgh + W_{air} = K_2 - K_1$$
Here, W_{air} = work done by air drag-
$mgh + W_{air} = K_2 - K_1$
or $W_{air} = K_2 - K_1 - mgh$
or $W_{air} = \frac{1}{2}m(1.1\sqrt{gH})^2 - mgh$
or $W_{air} = 0.605 \, mgH - mgh$
$\Rightarrow W_{air} = -0.395 \, mgH$

20. **(A)** Since, the springs are in parallel, therefore, the effective spring constant of the combination $k = k_1 + k_2 = 225 \, N/m$

By conservation of mechanical energy, we have,
PE of spring block system at displaced position
= KE of the block at mean position
i.e., $\frac{1}{2}kx^2 = \frac{1}{2}mv^2$

or $\frac{1}{2}(225)(0.1)^2 = \frac{1}{2}(2)v^2$
$\Rightarrow v^2 = 1.125 \Rightarrow v = 1.06 \, m/s$
Therefore, the velocity of the block at mean position will be, $v = 1.06 \, m/s$.

21. For (A): Initial acceleration
$$= \frac{k\left(\frac{3\mu mg}{k}\right) - \mu mg}{m} = 2\mu g$$

For (B, C)

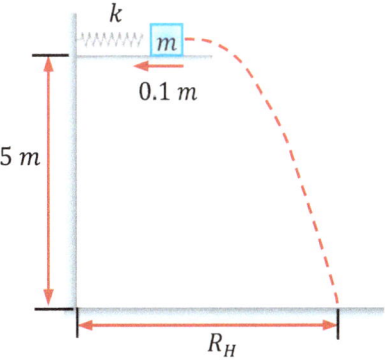

Therefore, maximum compression
$= \frac{2\mu mg}{k} - \frac{\mu m}{k} = \frac{\mu mg}{k}$
and minimum compression $= 0$
For (D): At maximum speed $F_{net} = 0$ so by using work energy theorem,
$\frac{1}{2}mv^2 = \frac{1}{2}k\left(\frac{3\mu mg}{k}\right)^2 - \frac{1}{2}k\left(\frac{\mu mg}{k}\right)^2 - \mu mg\left(\frac{2\mu mg}{k}\right)$
$\Rightarrow v = 2\mu g\sqrt{m/k}$

22. **(B)** By conservation of mechanical energy, we have,
∴ $\frac{1}{2}kx^2 = \frac{1}{2}mv^2$
∴ $x = v\sqrt{\frac{m}{k}} = 4\sqrt{\frac{16}{100}} = 1.6 \, m$

23. **APPROACH** First, we have to calculate the velocity (v) of the block when it leaves the platform by applying the principle of conservation of mechanical energy between the fully compressed position of spring and when the block is just about to leave the platform. This velocity will be horizontal. After leaving the platform, the motion of the block will be under gravity. Therefore, consider the vertical motion of the block and find the time taken (t) by the block to reach the ground. Now, we can find the horizontal range by using the equation of horizontal motion under gravity, i.e., $R = v \times t$

SOLUTION Applying the principle of conservation of mechanical energy between the fully compressed position of spring and when the block is just about to leave the platform-
$$\frac{1}{2}kx^2 = \frac{1}{2}mv^2$$

or $v = \sqrt{\frac{kx^2}{m}} = \sqrt{\frac{100 \times 0.01}{0.5}} = \sqrt{2}$ m/s

Now, after leaving the platform, the motion of the block is under gravity.
Set the origin at the ground, just below the point of projection of the block on the platform, we have
$v_{0y} = 0, a_y = -g, y_0 = +5m, y = 0, t = ?$
Applying the equation of motion,
$$y = y_0 + v_0 t + \frac{1}{2} a_y t^2,\text{ we get}$$
$$0 = 5 + 0 - \frac{1}{2} g t^2$$
or $t = \sqrt{\frac{5 \times 2}{g}} = \sqrt{\frac{5 \times 2}{10}} = 1\,s$

Now, consider the horizontal motion of the block,
Maximum horizontal range, $R_H = v \times t = \sqrt{2} \times 1$
or $R_H = 1.414\,m$
Thus, the block will hit the ground at $1.414\,m$ from a point $5m$ below the spring.
Therefore, option A is the correct choice.

24. (A) *By conservation of mechanical energy, we have-*
$$\frac{1}{2} m v_0^2 = \frac{1}{2} k x^2$$
or $x = \sqrt{\frac{m v_0^2}{k}} = v_0 \sqrt{\frac{m}{k}}$

∴ $F_{max} = kx = \sqrt{km} v_0$

25. The power is given by-
$$P = \vec{F}.\vec{v} = Fv$$
(Since, in a plane road, force applied by bus and its velocity, both are in same direction)
To overcome the resistive force R, the bus has to apply force $F = R$.
Therefore, the power at this instant,
$$P = Rv$$
Now, if v_{max} is the maximum velocity, then
$$P = R v_{max}$$
or $v_{max} = \frac{P}{R}$
When, $v = \frac{v_{max}}{2}$, then the force by which engine pulls the bus is
$F = \frac{P}{v} = \frac{R v_{max}}{v_{max}/2} = 2R$
Now, by Newton's second law, we have
$\sum F = ma$
i.e., $F - R = ma$
or $2R - R = ma$
or $a = \frac{R}{m}$

26. (D) We can consider the whole mass of the rod at its centre of mass.

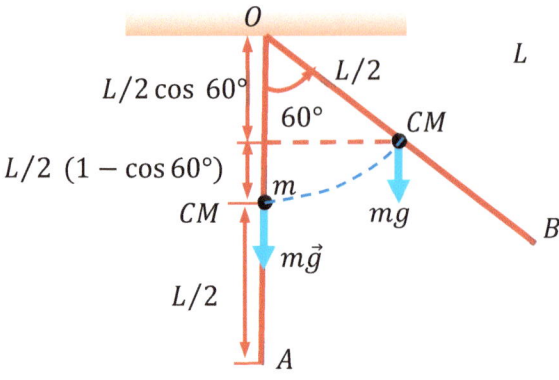

∴ Work done $= mgh$
$= mg \left(\frac{L}{2} - \frac{L}{2} \cos 60°\right)$
$= mgL \left(\frac{1}{2} - \frac{1}{4}\right) = \frac{mgL}{4}$

27. (B) From the given figure, we have $\sin\theta = \frac{3}{5}$ and $\cos\theta = \frac{4}{5}$

F.B.D. of the block

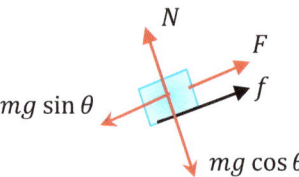

In above FBD, $F \to$ force by the man,
$f = \mu_k N = \mu\,mg \cos\theta \to$ frictional force
$N = mg \cos\theta \to$ normal reaction of the surface
& $mg \to$ gravitational force
Since block slides with constant speed, hence
$mg \sin\theta = F + f$
$\Rightarrow F = mg \sin\theta - f$
$= 10 \times 10 \times \frac{3}{5} - 0.1 \times 10 \times 10 \times \frac{4}{5}$
$= 60 - 8 = 52\,N$
$W_m = \vec{F}.\vec{s} = Fs \cos 180° = -Fs$
Here $F = 52\,N$, and $s = 5\,m$.
$\Rightarrow W_m = -52 \times 5\,J = -260\,J$

28. (A) $W_{gravity} = mgs \sin\theta$
$= 10 \times 10 \times 5 \times \frac{3}{5}\,J = 300\,J$

29. (D) $W_{surface} = W_N + W_{friction}$
$= 0 + fs \cos 180°$
$= 0 - \mu mg \cos\theta\,s$
$= -0.1 \times 10 \times 10 \times \left(\frac{4}{5}\right) \times 5\,J$
$= -40\,J$

30. (A) Work done by the resultant force is given by
$W = W_m + W_g + W_N + W_f$
$= -260\,J + 300\,J + 0 - 40\,J = 0$

31. (A) Since net work done by all the force is zero, Therefore, from work energy theorem, the change in $KE = 0$.

32. (A) $W = \vec{F}.\vec{r} = (3i + 4j + 5k).(3i + 4j + 6k)$

$= 9 + 16 + 30 = 55J$

33. APPROACH In this case, the only force on the projectile is downward gravitational pull, whereas its displacement is horizontal.

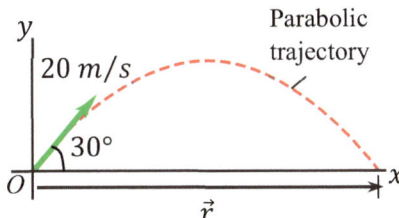
Parabolic trajectory

$\vec{F} = -mg\hat{j}, \vec{r} = r\hat{i}$

SOLUTION (A) Work $= (mg\,\hat{j}).(r\,\hat{i}) = 0$.

34. (A) $x = 3t^2$
Velocity $v = \frac{dx}{dt} = 6t\ m/s$
Acceleration $a = \frac{dv}{dt} = 6 m/s^2$
Force acting on particle $= ma = 1 \times 6 = 6N$
Therefore, force in vector form, $\vec{F} = (6N)\hat{i}$
Displacement $\vec{s} = (2\ m)\hat{i}$
Here, force is constant, therefore work done
work $W = \vec{F}.\vec{s} = (6N)\hat{i}.(2\ m)\hat{i} = 12J$

35. (D) When it hits the ground the net work done is zero. So, $\vec{P} = \frac{W}{T} = \frac{0}{T} = 0$.

36. (A) $a = \frac{4-0}{40} ms^{-2} = \frac{1}{10} ms^{-2}$
Velocity at the end of 8 second $= 0 + \frac{1}{10} \times 8 = 0.8 ms^{-1}$
Required power $= (ma)\,v = 8 \times \frac{1}{10} \times 8$
$= 0.8W = 0.64W$

37. (A) Given that power supplied by engine,
$P = c$, constant
$\therefore\quad Fv = c$
$\Rightarrow F = \frac{P}{v}$
At the time of maximum velocity $F = r$. I.e., the net force on load $= 0$
$\therefore\quad v_{max} = \frac{P}{r}$
Again, from, $F = \frac{P}{v}$, we have
$m.\frac{dv}{dt} = \frac{P}{v}$
or $\quad v\,dv = \frac{P}{m} dt$
or $\quad \int_0^{v_{max}/2} v.dv = \left(\frac{P}{m}\right) \int_0^t dt$
or $\quad \left[\frac{v^2}{2}\right]_0^{v_{max}/2} = \left(\frac{P}{m}\right)[t]_0^t$
or $\quad \frac{v_{max}^2}{8} = \left(\frac{P}{m}\right) t$

or $\quad t = \frac{m}{P}\frac{(P/r)^2}{8} = \frac{Pm}{8r^2}$

38. (C) Since, the block connected with spring performs simple harmonic motion, therefore the displacement of block from its mean position can be written as-

$\quad\quad x = A \sin \omega t \quad\quad\quad$ ... (1)

Here, A is the amplitude of the motion.
On differentiating both sides, we get-
$\quad$ Velocity, $v = \frac{dx}{dt} = A\omega \cos \omega t \quad$ ... (2)
And acceleration, $a = -A\omega^2 \sin \omega t \quad$ ... (3)
At $t = 0$, $v = A\omega \cos 0 = A\omega = v_0$ (given)
$\therefore\quad A = \frac{v_0}{\omega}$
Substituting this value of A, in (2), and in (3), we get
$\quad\quad v = v_0 \cos \omega t \quad\quad\quad$ ... (4)
and $\quad a = -\frac{v_0}{\omega} \omega^2 \sin \omega t$
or $\quad a = -v_0 \omega \sin \omega t \quad\quad$ ... (5)
Now, force $F = ma = -m v_0 \omega \sin \omega t$
Here, $-$ sign shows that this force is directed towards mean position.
When block is moving away from the mean position, the spring force and velocity both will be in opposite direction.
In this case, instantaneous power
$\quad\quad P = \vec{F}.\vec{v} = Fv \cos 180° = -Fv$
$\quad\quad = -(mv_0\omega \sin \omega t)(v_0 \cos \omega t)$
$\quad\quad = \frac{mv_0^2 \omega}{2} (2 \sin \omega t . \cos \omega t)$
$\quad\quad = \frac{mv_0^2 \omega}{2} \sin 2\omega t$
$\therefore$ Maximum power, $P_{max} = \frac{mv_0^2 \omega}{2} (\sin 2\omega t)_{max}$
or $\quad P_{max} = \frac{mv_0^2 \omega}{2} \quad [\because (\sin 2\omega t)_{max} - 1]$
Now, corresponding to maximum power,
$\quad (\sin 2\omega t)_{max} = \sin \left(n\pi + (-1)^n \frac{\pi}{2}\right)$
$\therefore\quad 2\omega t = n\pi + (-1)^n \frac{\pi}{2}$
here, $n = 0,1,2,...$
or $\quad t = \frac{\pi}{4\omega}, \frac{3\pi}{4\omega}$ etc.

39. (A) →(T); (B) →(P); (C) →(Q); (D) →(R)
SOLUTION

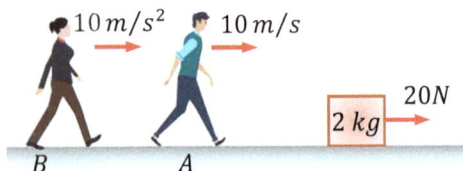

For (A): Work energy theorem is applicable in all reference frames.

For (B) : with respect to ground: At $t = 0$, $v_0 = 0$ and $t = 1\,s$, $v = at = \left(\frac{20}{2}\right)(1) = 10\,\text{m/s} =$

Work done = change in kinetic energy = $\frac{1}{2}(2)(10)^2 - \frac{1}{2}(2)(0)^2 = 100\,J$

For (C): with respect to observer A: Initial velocity = $0 - 10 = -10\,m/s$, Final velocity = $10 - 10 = 0$
Work done = $\frac{1}{2}(2)(0)^2 - \frac{1}{2}(2)(-10)^2 = -100\,J$

For (D): with respect to observer B: Initial velocity = $0 - 0 = 0$
Final velocity = $10 - 10 = 0$; Work done = 0

40. (B) By work energy theorem, we have-
$$\omega_{spring} + \omega_{friction} = 0 - \frac{1}{2}mv_0^2$$
$$\Rightarrow -\frac{1}{2}kx^2 - \mu mgx = -\frac{1}{2}mv_0^2$$
$$\Rightarrow kx^2 + 2\mu mgx - mv_0^2 = 0$$

or $$x = \frac{-2\mu mg \pm \sqrt{4\mu^2 m^2 g^2 + 4kmv_0^2}}{2k}$$

or $$x = \frac{\mu mg + \sqrt{\mu^2 m^2 g^2 + 4kmv_0^2}}{k}$$

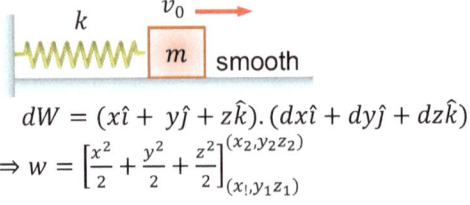

So, work done by spring force = $-\frac{1}{2}km^2$

41. (A) Let a particle on which force $\vec{F}$ is acting moving from
(x_1, y_1, t_1) to (x_2, y_2, t_2) then
$dW = \vec{F}.d\vec{r}$ and $d\vec{r} = dx\hat{\imath} + dy\hat{\jmath} + dz\hat{k}$

$dW = (x\hat{\imath} + y\hat{\jmath} + z\hat{k}).(dx\hat{\imath} + dy\hat{\jmath} + dz\hat{k})$
$\Rightarrow w = \left[\frac{x^2}{2} + \frac{y^2}{2} + \frac{z^2}{2}\right]_{(x_1,y_1,z_1)}^{(x_2,y_2,z_2)}$

So, above work depend only on the coordinates of point (x_1, y_1, z_1) and (x_2, y_2, z_2). Therefore, force is conservative.

42. (C) $\frac{dv}{dt} = k^2rt^2 \Rightarrow v = \frac{k^2rt^3}{3}$
Centripetal force is always perpendicular to tangential velocity, therefore from $P = \vec{F}.\vec{v}$, it is clear that the power delivered by centripetal force will be zero. The tangential force $F_t = m\frac{dv}{dt}$
Power delivered by tangential force,
$$P = \vec{F}.\vec{v} = \left(m\frac{dv}{dt}\right).v$$
$$= mk^2rt^2 \cdot \frac{k^2rt^3}{3} = \frac{mk^4r^2t^5}{3}$$

43. (A) By work energy theorem,

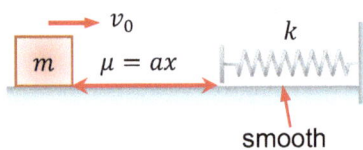

$W_c + W_{nc} = K_2 - K_1$
or $\quad W_{spring} + W_{friction} = K_2 - K_1 \quad \ldots (1)$

$K_1 = \frac{1}{2}mv_0^2$
$K_2 = 0$
$W_{spring} = $ −change in PE $= -\frac{1}{2}kx^2$
$W_{spring} = $ −change in PE
$W_{friction} = -\int_0^L axmg\,dx = -\frac{amgL^2}{2}$
Putting above values in Eq. 1, we get-
$-\frac{1}{2}kx^2 - \frac{am}{2} = 0 - \frac{1}{2}mv_0^2$
or $\frac{1}{2}mv_0^2 = \frac{1}{2}kx^2 + \frac{amgL^2}{2}$
$\Rightarrow x = \sqrt{\frac{mv_0^2 - amgL^2}{k}}$

44. (C) Total mechanical energy = kinetic energy + potential energy = $15 + [32 - 6(3) + 14]$
$= 15 + 5 = 20\,J$

45. (B) At maximum speed (i.e. maximum kinetic energy), potential energy is minimum
$U = x^2 - 6x + 14 = 5 + (x - 3)^2$
which is minimum at $x = 3\,m$ so $U_{min} = 5J$.
Therefore, $K_{max} = 20 - 5 = 15J$
$\Rightarrow \frac{1}{2}mv_{max}^2 = 15$
$\Rightarrow v_{max} = \sqrt{30}\,m/s$

46. (C) For particle
$K \geq 0 \Rightarrow E - U \geq 0$
$\Rightarrow 20 - (5 + (x - 3)^2] \geq 0$
$\Rightarrow (x - 3)^2 \leq 15$
$\Rightarrow x - 3 \leq \sqrt{15} \Rightarrow x \leq 3 + \sqrt{15}$

47. (C) $P = Fv = mav$

$$= mv\frac{dv}{dt} = 4t^3 - 5t + 2$$

$$K = \frac{mv^2}{2} = t^4 - \frac{5t^2}{2} + 2t$$

$K_1 = 16 - 10 + 4 = 10\,J$ (At $t = 2$ sec.)
$K_2 = 256 - 40 + 8 = 224\,J$ (At $t = 4$ sec.)

So, change in K.E. will be $224\,J - 10\,J = 214\,J$.

48. (C) $P = \frac{W}{t}$

$$t = \frac{W}{P} = \frac{mgh}{P} = \frac{200 \times 10 \times 40}{10 \times 100} = 8s$$

49. (A) $W = F.d = Fd\cos\theta$

$$\theta = 90°$$

then $W = 0$

50. (B) Mass of water pumped per second

$m = \frac{2400}{60} = 40\,kg$

$v = 3\,m/s$

Kinetic energy of water coming out per second

$$= \frac{1}{2}mv^2 = \frac{1}{2} \times 40 \times 3 \times 3 = 180\,J$$

From work energy theorem, we have

$$W = K_2 - K_1 = 180\,J$$

∴ Average power of pump $\frac{W}{t} = 180\,J/s = 180\,W$

33.9. MULTIPLE CHOICE ASSIGNMENTS

33.9.1. LEVEL 1

Q. No.	1	2	3	4	5	6	7	8	9
Ans	C	D	B	C	B	B	B	C	D
Q. No.	10	11	12	13	14	15	16	17	18
Ans	A	B	C	A	B	A	D	C	A
Q. No.	19	20	21	22	23	24	25	26	27
Ans	B	C	A	B	A	B	C	D	C

Q. No.	28	29	30	31	32			
Ans	D	A	B	C	D			

33.9.2. LEVEL 2

Q. No.	1	2	3	4	5	6	7	8	9
Ans	A	A	C	D	B	B	A	C	A
Q. No.	10	11	12	13	14	15	16	17	
Ans	B, D	C	A	B	B	A	C	B	

33.9.3. LEVEL 3

Q. No.	1	2	3	4	5	6	7	8	9
Ans	A	C	D	C	B	B	D	A	B
Q. No.	10	11	12	13	14	15	16	17	18
Ans	A	A	B	C	D	A	D	A	C
Q. No	19				20				
Ans	A→Q, B→S, C→R				A→T, B→P, C→S, D→Q				

33.9.4. LEVEL 4

33.9.4.1. SECTION A

Q. No.	1	2	3	4	5	6	7	8	9
Ans	B	A	B	B	A	B	D	B	C
Q. No.	10	11	12	13	14	15	16	17	18
Ans	A	C	B	A	D	A	D	A	D
Q. No.	19	20	21	22	23	24	25	26	27
Ans	B	C	A	D	D				

33.9.4.2. SECTION B

Q. No.	1	2	3	4	5	6	7	8	9
Ans	D	D	D	B	D	B	C	C	D
Q. No.	10	11	12	13	14		15		16
Ans	D	4	5	5	A, B, D		30.00		0.75J

www.ingramcontent.com/pod-product-compliance
Lightning Source LLC
Chambersburg PA
CBHW051153220526
45473CB00003B/753